MEDIA AND COMMUNICATIONS - TECHNOLOGIES, POLICIES AND CHALLENGES

SOCIAL NETWORKING

RECENT TRENDS, EMERGING ISSUES AND FUTURE OUTLOOK

MEDIA AND COMMUNICATIONS - TECHNOLOGIES, POLICIES AND CHALLENGES

Additional books in this series can be found on Nova's website under the Series tab.

Additional e-books in this series can be found on Nova's website under the e-books tab.

COMPUTER NETWORKS

Additional books in this series can be found on Nova's website under the Series tab.

Additional e-books in this series can be found on Nova's website under the e-books tab.

MEDIA AND COMMUNICATIONS - TECHNOLOGIES, POLICIES AND CHALLENGES

SOCIAL NETWORKING

RECENT TRENDS, EMERGING ISSUES AND FUTURE OUTLOOK

XIN MING TU
ANN MARIE WHITE
AND
NAIJI LU
EDITORS

New York

For permission to use material from this book please contact us:
Telephone 631-231-7269; Fax 631-231-8175
Web Site: http://www.novapublishers.com

Library of Congress Cataloging-in-Publication Data

Social networking : recent trends, emerging issues and future outlook / editors, Xin Ming Tu, Ann Marie White and Naiji Lu.
pages cm
Includes bibliographical references and index.
ISBN: 978-1-62808-529-7 (hardcover)
1. Social networks--Research. 2. Online social networks--Research. I. Tu, Xin M. II. White, Ann Marie. III. Lu, Naiji, Ph.D.
HM741.S6346 2013
302.30285--dc23

2013026853

Published by Nova Science Publishers, Inc. †New York

CONTENTS

Preface		**vii**
Chapter 1	The Effect of Filtering on Animal Networks *Nienke Alberts, Stuart Semple and Julia Lehmann*	**1**
Chapter 2	Hash Tags, Status Updates and Revolutions: A Comparative Analysis of Social Networking in Political Mobilization *Patience Akpan-Obong and Mary Jane C. Parmentier*	**21**
Chapter 3	Relationships between Personality and Interactions in Facebook *Fabio Celli and Luca Polonio*	**41**
Chapter 4	Resilience to Climate and Demographic Change: The Importance of Social Networks *Kaberi Gayen and Robert Raeside*	**55**
Chapter 5	Social Networking Services and Analysis: The Third Revolution in Behavioral Research? *Christopher M. Homan and Vincent M. B. Silenzio*	**73**
Chapter 6	Social Networking: Addressing an Unmet Need in the Young Haemophilia Population *Kate Khair, Mike Holland and Shawn Carrington*	**101**
Chapter 7	My Best Potential Friend in a Social Network *Francisco Moreno, Andrés González and Andrés Valencia*	**113**
Chapter 8	The Role of Opinion Leaders and Internet Marketing through Social Networking Websites *Viju Raghupathi and Joshua Fogel*	**125**
Chapter 9	Social Networks and the Job Search: A Focus on People Who are Asked to Provide Job Assistance *Lindsey B. Trimble, Julie A. Kmec and Steve McDonald*	**139**

Chapter 10 Implications of Social Network Endogeneity: From Statistical to Causal Inferences **167**
N. Lu, A. M. White, P. Wu, H. He, J. Hu, C. Feng and X. M. Tu

Index **185**

PREFACE

"Social networks" is no longer a term solely of the academe. A Google search of the term "social networks" at the time of writing this Preface yielded over 117,000,000 hits. Searching for "network analysis" yielded 6,460,000 hits. This is a field of study drawing upon graph theory to represent relationships between objects (e.g., websites) and how these form functional systems. "Social network analysis," the counterpart of network science focused on human interactions, yielded 939,000 or close to one million hits. Arguably, the advent of social media created a cultural meme that carries network concepts forward to general publics.

Social network analysis is a robust means to measure and map interactive systems. An approach to dynamically represent interactions among people and how these are organized – this method can capture ties within a system (e.g., between people) and between systems (e.g., between humans and other species or between humans and their environments). Due to inherent flexibility, network analytic approaches are now ubiquitous across any of a number of theoretical and empirical efforts in fields too numerous to list.

Social network analysis (SNA) focuses on social relationships (e.g., friendship) diagramed as nodes (points) and links (ties or edges between points). SNA examines features or changes to a social system illuminated from interactions and how these change over time (e.g., spread of disease). Social network analysis can measure and map "connectedness" or "flows" (e.g., information, resources, etc.) both within and across individuals, groups and organizations, and can locate these in a virtual- or geo-temporal space – yielding rich mixed methods possibilities (e.g., tying together joint analysis of physical or place-based attributes with social network attributes).

A notion that originated in sociology over a century ago, a social network is, at its heart, a social science theoretical concept that has widely spread to become an interdisciplinary application. As a method, it has yielded highly influential studies that generate implications beyond qualitative descriptions. For instance, Christakis and Fowler's 2007 article in the *New England Journal of Medicine* that drew upon Framingham Heart Study data, helped mobilize obesity as a public health priority. While in essence caused by behaviors not pathogens, the use of network analyses helped characterize obesity as contagious, or a spreading disease, and effectively conveyed the notion of an 'obesity epidemic' to rally greater public concerns.

Relationships between actors in a social group comprise a general mechanism underpinning any of a number of key accumulations or traits attributed to a social context or system – such as capital, efficiency or optimization, evolution, ecology, and spreading of information or ideas. As the referent of "social networks" (SN) has spread in academic fields as well as general everyday use, so have methods for illuminating and testing its properties and applications. From the early sociogram methods in sociology, to the advent of sophisticated visualization techniques, and most recently to its application in big data mining, the pace of innovation in SN methods and applications of concepts continues unfettered.

The growth and application of methods for studying social networks continues to burgeon in many fields. Given the scientific advancements in the study of networks over recent decades, highlights of this approach are now easily found among, and generated by, business, health, environment, computer science, statistics, economics and biology applications. Together, social network concepts and related analytic methods have propelled many fields towards greater understanding of critical phenomena such as infection, popularity or influence (e.g., page rank), and efficiency or capacity (e.g., transportation or trade).

However, applications of methodological advancements in many fields lag behind the pace of innovation. And fundamental limitations to surmount, to give this method greater explanatory (as compared to descriptive) power, have remained. Thus, this book is designed to promote wider reading about social network advancements across fields to accelerate the pace of interdisciplinary explorations and potential new discoveries.

The extensiveness and spread of SN theory and SNA's use and application suggests that emerging work in the many disciplines applying this field can be culled and encapsulated to advance research and training widely. Cross-pollination with other fields further astride in SNA is needed as many sciences (e.g., child development) have omitted foundational concepts in basic preparation of their future scientists. Our main objective for this volume is thus to share a wide set of field-specific insights that have the potential to advance other fields embracing similar foci.

We do so by first discussing conceptual issues in SNA, including the potential for applications of SN concepts yet to be addressed or encapsulated for wide dissemination within a particular field of study. For instance, a chapter in this volume poses answers to the question of what form SNA takes across the "observational-experimental" research design "continuum," an important topic in public health sciences. Basic questions about the power or utility of SN concepts are advanced via literature reviews (e.g., the role of opinion leaders in purchase behaviors).

Our volume also turns to methodological applications and advances in SNA. Currently, computer scientists increasingly seek to bring statistical expertise to bear in network analysis due to a host of unknowns to surmount in SNA. For instance, how do different strategies to identify a group affect conclusions about membership? How can we address a forgotten, but omnipresent, dimension of network datasets for valid statistical and causal inference (violation of independence assumption due to shared group membership) with SN datasets? Explorations to generate potential solutions or applications are presented.

Our goal for this book is to capture – across a wide range of fields – how emerging issues in the application of SN theory and SNA are being addressed. How to press forward past edges of our knowledge are illuminated as such a diverse set of authors' disciplinary expertise are brought to SNA. Each chapter selected illuminates new trends or applications that may have wide potential impact in other disciplines. Conceptual advances (e.g., applying the concepts of social networks such as peer influence on disease management and treatment adherence) and novel analytic approaches for studying properties of social networks are both highlighted.

These chapters convey that many frontiers remain for the study and application of social networks. Similar to a Google search described above, our call for papers shows that academic fields are at varying different stages of application of SNA. In our preparations for this book, we thus sought a call for papers that in fact yielded great diversity. We decided early on, that it is important for any book on SNA to share insights widely across fields.

Taken together, a picture emerges from the selected chapters that much work remains, work that can be cross-fertilized through multidisciplinary team building. The potential for this kind of team assembly grows as SN and SNA ideas spread and find fertile ground within respective fields. This volume demonstrates that with such a base, the next generation of SN concepts and methods can be propelled if the coalescing of interdisciplinary teams is fostered.

On a practical note, our efforts would not be in vain, if any reader "captures" an idea for use in their thinking or practice, or even if a reader seeks to connect to the work and commitment of these authors advancing SNA in each of their fields. Towards these ends – we invite readers to contact us as this volume "spreads" or influences yours or others attitudes, skills, behaviors, relationships and actions. We post our email contact information below. We look forward to hearing from readers about your use of this volume and working with you to grow this network of science.

Sincerely,

Naiji Lu
Department of Biostatistics and Computational Biology
Naiji_Lu@urmc.rochester.edu

Ann Marie White
AnnMarie_White@urmc.rochester.edu

Xin Tu
Xin_Tu@urmc.rochester.edu
Department of Biostatistics and Computational Biology
Rochester New York
University of Rochester

May 2013

In: Social Networking
Editors: X. M. Tu, A. M. White and N. Lu
ISBN: 978-1-62808-529-7

Chapter 1

THE EFFECT OF FILTERING ON ANIMAL NETWORKS

Nienke Alberts, Stuart Semple and Julia Lehmann
Centre for Research in Evolutionary and Environmental Anthropology,
University of Roehampton, London

ABSTRACT

The past few years have seen a surge in the use of social network analysis to study animal sociality. Because ties between animals are usually inferred from behavioural interactions or spatial proximity, animal social networks may contain ties that are due to chance rather than representing a true 'bond'. To help focus on relationships that are more likely to be biologically meaningful, networks are often filtered by removing all the ties that are under a certain cut-off value. Researchers have proposed various methods to determine the level of filtering; however, it is not clear how these different methods of filtering may alter network metrics and consequently how they may affect the conclusions that are drawn from subsequent analyses. We investigate the effect that five commonly used filtering methods have on standard network metrics. To this end, social networks were generated using association indices of a troop of wild olive baboons (*Papio anubis*). These networks were filtered (i) until the network structure was significantly different from random, (ii) by median association strength, (iii) by mean association strength, (iv) until the giant component was close to breaking up, (v) by including only preferential associations. Global network metrics, individual network positions, and the extent of substructuring were determined and compared across the five filtered networks and the unfiltered network. Our results show that while global network metrics and individual network positions are affected by different filtering methods in a relatively predictable way, the number of substructures that were found in networks was strongly influenced by the way filtering was done. These results show that the appropriate filtering method needs to be carefully considered, based on the nature of the biological questions being asked.

INTRODUCTION

The last decade has seen a marked increase in the use of social network analysis in the study of animal societies (Krause et al., 2009; Brent et al., 2011) across a variety of taxa, ranging from fish (e.g. Croft et al., 2004; Croft et al., 2006) to primates (e.g. Sueur and Petit 2008; Lehmann and Boesch 2009; Ramos-Fernández et al., 2009; Henkel et al., 2010; Lehmann and Ross 2011), from elephants (e.g. Wittemyer et al., 2005) to squirrels (e.g. Manno 2008). This approach has proven valuable in addressing a wide range of questions about animal societies. For example, social network analysis has been used to characterise the social structures and identify substructures in animals groups in order to make fine-grained comparisons between populations, species and across time, (Lusseau et al., 2006; Sundaresan et al., 2007; Kasper and Voelkl 2009) and to assess the effect of environmental factors on social structure (Wittemyer et al., 2005; Henzi et al., 2009; Alberts 2012). The social network approach is frequently used to study animal populations with high levels of fission-fusion dynamics, *i.e.* where groups frequently split and reform, and where social structure is therefore not readily apparent (Ramos-Fernández et al., 2006; Sundaresan et al., 2007; Wolf and Trillmich 2008; Ramos-Fernández et al., 2009). In such populations, social network analysis is often used to identify layers in the social structure that were hitherto unknown (Lusseau et al., 2006; Ramos-Fernández et al., 2006; Sundaresan et al., 2007; Wolf et al., 2007). The social network approach has also been used for the identification of stable social bonds between individuals, and investigation of how these bonds may benefit individuals (Croft et al., 2004; Lehmann and Boesch 2009; Lea et al., 2010; Brent et al., 2011). An important use of social network analysis is to investigate the role of individuals within networks (Mitani 1986; Flack et al., 2006; Sueur and Petit 2008; Ramos-Fernández et al., 2009; Henkel et al., 2010), and in particular to explore how individual characteristics, such as sex or age, may influence an individual's position in their social network (Blumstein et al., 2009; Ramos-Fernández et al., 2009; Lehmann and Ross 2011).

An important issue that has arisen from studies of animal social networks in the last decade is how relationships between individuals are defined, and what constitutes a 'tie'. In animal social networks, relationships between individuals are inferred, usually from behavioural interactions or from spatial proximity. For some behavioural interactions, such as grooming, inferring social relationships can be relatively straightforward, as individuals target particular group members, and may make a considerable investment of time in such an interaction. However, inferring relationships from other behaviours may result in social networks that contain ties that do not represent a true 'relationship', but rather are due to chance encounters. In particular, networks that are based on associations defined by 'the gambit of the group' (Whitehead and Dufault 1999), *i.e.* when individuals are assumed to be associated when they are found in the same group, are more likely to contain ties that are due to chance events (Croft et al., 2008). In addition, unlike human groups, in many taxa societies are relatively small and closed, so that there is a clearly demarcated group of individuals that belong to the network, and changes to the composition of the network only occur through demographic changes (*i.e.* births, deaths, emigrations, immigrations). Such societies may lead to social networks in which all, or the majority of individuals are interconnected (Jacobs and Petit 2011), making it impossible to differentiate between network metrics of different networks if weighted networks are not used. In such weighted metrics, the strengths of the relationships are indicated by the weight of ties. The development of network metrics that specifically take into account the weights of ties, such as those implemented in *tnet* (Opsahl 2009), has been an important advance in this field. Such weighted network metrics can, for example, help to differentiate between an animal that is very social, and thus has many strong ties, and an animal that has many weak relationships. Nevertheless, the majority of network metrics currently do not take into account the weights of ties. Thus, even when metrics are calculated in a weighted network, network metrics often only take into account whether ties are absent or present, and not the weights of ties.

To help differentiate between chance ties and real relationships, in other words to help focus on relationships that are more likely to be biologically meaningful, animal social networks are often filtered until non-random core elements remain (Croft et al., 2008). In these cases, a filter is applied to a network by removing all the ties that are under a certain cut-off value, which can in principle be set at any level (Croft et al., 2008; Sueur et al., 2011). While it is common to filter animal networks prior to analysis, there is no set method for determining the appropriate cut-off value that should be used (Croft et al., 2008), and several different approaches have been proposed. Methods include filtering the network until its structure, which is measured by certain network metrics as test statistics, is significantly different from a random network structure (Brent 2009; Alberts 2012), filtering by median or mean association strength (Croft et al., 2004; Croft et al., 2008), or filtering until the giant component of the network is close to breaking up into smaller components (Croft et al., 2008). Perhaps the most frequently used method for the filtering of animal networks is to include only preferential associations (Whitehead 1999; Lusseau 2003; Wittemyer et al., 2005; Lusseau et al., 2006; Williams and Lusseau 2006; Sundaresan et al., 2007; Manno 2008; Lehmann and Boesch 2009; Ramos-Fernández et al., 2009; Henkel et al., 2010). In those studies, an association is considered 'preferential' when individuals associate significantly more frequently than predicted if individuals associated with each other at random. The choice of filtering methods is usually based on the type of data, *i.e.* binary or weighted, used in the study as well as the research questions that are addressed. While filtering of networks may be appropriate in studies that focus on the social relationships of individuals, it may not be for studies that focus on the transmission of disease or parasites (e.g. Corner et al., 2003; Cross et al., 2004; Godfrey et al., 2009) in which rare chance events may be very important. Researchers of animal social networks frequently use global network metrics to indicate the structure or qualities of the network to draw conclusions about their study groups. Additionally, researchers often determine individual network positions, and how these can be predicted by individual characteristics. Finally, social network analysis has been popular in helping to determine the level of substructuring of the network. Currently it is not known how each of these levels of investigation are affected by the various methods of filtration, and consequently how the filtration method used may affect the conclusions that are drawn from these metrics.

In this chapter we investigate the effect of different filtering methods on commonly used network metrics, using the association network of a troop of wild olive baboons as a case study. Global network metrics, individual network positions, and the extent of network substructuring were compared across the unfiltered network and networks that were filtered: (i) until the network structure was significantly different from random, (ii) by median association strength, (iii) by mean association strength, (iv) until the giant component was close to breaking up, and (v) by including only preferential associations.

METHODS

Data Collection and Calculation of the Association Index

Data were collected on a troop of wild olive baboons (*Papio anubis*) in Gashaka-Gumti National Park (GGNP), north-eastern Nigeria, over a one-year period from March 2009-March 2010. GGNP is on the margin of the distribution of baboons, and is a somewhat unusual study site for baboons, as it includes large areas of rainforest and is the wettest of all the baboon study sites (Higham et al., 2009). The 'Kwano' study troop has been studied continuously since 2000 (Sommer and Ross 2011). For the current study, baboons were fully habituated to human observers and could be followed at a 2-6m distance. All baboons were individually recognised by the observers. The Kwano troop had a mean group size of 34 individuals (range 31-37) during the study period; however, only adults and subadults that were present for the entire study period were included in the networks (*i.e.* 16 individuals). The Kwano troop forms a 'closed' social system, in which members are clearly identifiable, and membership is highly stable. Nevertheless, fission-fusion dynamics have been observed in this troop (Alberts 2012); therefore, individuals do not always associate simultaneously with all troop members, but instead the troop temporarily splits into smaller subgroups.

Data were collected each day over an eight-hour period (*i.e.* 06:00 – 14:00 or between 10:00 – 18:00). Instantaneous sampling of the group, or scan sampling (Altmann 1974), was conducted every hour. During scan sampling, the identity of each baboon in sight was recorded at a preselected moment in time (*i.e.* every hour). Individuals that were seen together in a scan were considered to be associated. For five minutes before each scan, researchers walked around the area to locate baboons. The definition of an association used here is thus broader than an association based on individuals being in visual contact. The method used here may be a more appropriate estimation of associations at this site; the terrain at GGNP is very uneven and large parts are forested, and therefore using a purely visual definition of associations may underestimate the number of individuals in a subgroup. Each day around eight scans were collected, making a total of 467 scans.

Individuals were considered to be in association if they were both observed in a scan. Scan data were used to calculate the Twice Weight Index (TWI) in SOCPROG 2.4 (Whitehead 2009). The TWI was calculated for each dyad as follows:

$$TWI = \frac{X}{X + Y_a + Y_b}$$

Where X is the number of times *a* and *b* were seen together, Y_a the number of times *a* was seen but not *b*, and Y_b the number of time *b* was seen but not *a* (Cairns and Schwager 1987).

Filtering of Networks

The TWIs were used to create six networks: one unfiltered, and five filtered. In the unfiltered network all TWIs were used directly. The second network was filtered until network structure was significantly different from random. To this end, the observed network was filtered by increments of 0.01 and dichotomised until the mean clustering coefficient and the mean geodesic were significantly different from random. These two metrics were calculated for the observed network and compared to the distribution of the test statistics for 50 Erdös-Rényi random graphs (Erdös and Rényi 1959). The mean clustering coefficient and mean geodesic were chosen as test statistics because they provide a measure of the average cohesion of a network, and are complementary; the mean geodesic is a global network measure, in that it considers paths over the whole network, whereas the clustering coefficient focuses more on local structures (Croft et al., 2008). Network structure was found to be significantly different from a random structure at a filtration level of 0.170. This filter was applied to the weighted network; thus associations weaker than 0.170 (*i.e.* TWI < 0.170) were excluded. The third network was filtered by median association strength (0.220), and the fourth by mean association strength (0.224) (*i.e.* TWIs less than these values were excluded). The fifth network was filtered until its giant component was close to breaking up into smaller components. The giant component, or largest connected component, is the component in the network that contains the majority of nodes. The mean degree of a network is often close to 1 when the giant component is close to breaking up (Croft et al., 2008), and drops below 1 when the component fragments. Individuals dropping out of the network were not considered separate components, and thus only when a second component of at least two nodes was observed, was the network considered to have fragmented into multiple components. To determine the appropriate level of filtering using this method, the unfiltered network was visualised using NetDraw 2.089 (Borgatti 2002). The network was then filtered at increments of 0.0001, and was visualised again at each stage. These steps were repeated until the giant component was observed to break up (*i.e.* when more than one component was observed), and then the previous increment was used as a cut-off value. The cut-off value for this method was set at TWI < 0.340. Finally, the sixth network was filtered to include only preferential associations. Preferentially associating dyads are pairs of baboons that were observed to be in association significantly more frequently than expected by chance given the number of times those individuals were observed. We tested for preferential associations using SOCPROG 2.4 (Whitehead 2009). In this procedure, observed TWIs are compared to the distribution of TWIs of randomised data sets, in which data are randomised using a modification of the methods of Manly (1995) and Bejder, Fletcher, and Bräger (1998) (20,000 permutations), keeping group size and the number of times each individual was observed constant (Whitehead 1999). Dyads that were observed to associate significantly *more* than expected by chance were included in the weighted network.

Calculation of Global Metrics, Individuals Network Positions, and Substructures

All network metrics were calculated in UCINET (Borgatti et al., 2002). To test the effect of filtering on global network structure, we compared eight network metrics that are commonly used in the study of animal social networks: the binary and weighted density, binary mean degree, the largest connected component, the mean geodesic, the diameter, the mean clustering coefficient, and the weighted network centralisation (see Table 1 for definitions) across the six networks. These eight metrics are frequently used by researchers of animal networks to characterise the structure of relationships in a group. Networks were dichotomised to calculate the binary density and the binary mean degree. While all other metrics were calculated in weighted networks, only the weighted density and weighted network centralisation take the weights of the ties into account specifically. When a network had more than one component, the diameter was calculated as the largest geodesic observed in any component, whereas the clustering coefficient calculations only include individuals with more than one tie.

Table 1. Definitions of global network metrics (adapted from Wasserman and Faust 1994; Croft et al., 2008; Opsahl and Panzarasa 2009).

Global network metrics	
Binary density	The density of a network is a measure of the number of ties in a network, and indicates the level of cohesion. It indicates the number of ties in relation to the possible number of ties. For an undirected network: $$\Delta = \frac{E}{n(n-1)/2}$$ Where E is the number of ties in the network, and n the number of nodes. The value of Δ ranges between 0 (empty) to 1 (completely connected).
Weighted density	The weighted density of a network is a measure of the average weight of ties across all possible ties, and indicates how strongly connected a network is. $$\Delta_\omega = \frac{E_\omega}{n(n-1)/2}$$ Where $E\ \omega$ is the sum of the values of all ties, and n the number of nodes.
Binary mean degree	The mean degree is a measure of how well connected a network is. It indicates how many ties nodes in the network have on average. $$k = \frac{1}{n}\sum_i k_i$$ Where k_i is the number of nodes i is connected to. Higher values indicate that on average individuals have more ties.
Largest connected component	The largest connected component is the size of the largest group of nodes that are all reachable from each other.
Mean geodesic	The average shortest path length is a measure of how close, on average, two individuals are to each other in the network. It indicates the shortest path from a node to all other nodes in the network. $$L = \frac{1}{n}\sum_{i=1}^{n} d_{ij}$$ Where n is the number of nodes in the network, and d is the shortest distance between node i and j. Larger values indicate a greater distance between individuals and thus that relationships are less direct.
Diameter	The network diameter is the largest of the shortest paths between individuals in a network. This gives an indication of how 'wide' a network is, in other words, the maximum distance between nodes. Higher values indicate individuals in the network may be more distant from each other.

Global network metrics	
Mean clustering coefficient	The mean clustering coefficient is a measure of the cliquishness of a network. It indicates the average proportion of ego's neighbours that are also connected to each other. $$C = \frac{1}{n}\sum_{i=1}^{n}\frac{2t_i}{k_i(k_i-1)}$$ Where t_i is the number of triangles of which node i is part, and k is the number of nodes i is connected to. The clustering coefficient ranges between 0-1, with large values indicating a large proportion of a node's neighbours also have ties between themselves (clustered).
Weighted network centralisation	The weighted network centralisation is a measure of how evenly ties are distributed over individuals in the network. It indicates the differences between the largest individual centrality score and the scores of all the other individuals in the network, and is normalised by maximum possible difference. $$C_D = \frac{\sum_{i=1}^{n}[C_D(n*) - C_D(n_i)]}{max\sum_{i=1}^{n}[C_D(n*) - C_D(n_i)]}$$ Where $C_D(n_i)$ is the centrality score for node i, and $C_D(n*)$ is the largest observed value. Low values indicate ties are equally distributed over individuals, high values indicate that a few individuals have most of the ties.

Next, we calculated four standard individual centrality measures frequently used in studies of animal social networks: degree centrality, betweeness centrality, closeness centrality and eigenvector centrality (see Table 2 for definitions). These centrality measures are frequently used by researchers of animal networks to identify animals that play important roles in their networks, and to draw conclusions about the characteristics of animals (*e.g.* age or sex) that may influence the role of individuals in their networks. Because centrality measures were calculated in weighted networks, the degree centrality and the eigenvector centrality use the sums of the values of an individual's ties. While betweeness and closeness centrality are distance-based measures, these do not take into account the weights of the ties, despite being calculated on a weighted network. To determine if individuals had similar network positions in networks that were filtered by different methods, and thus to assess the effect of filtering on the conclusions drawn about the roles of individuals in their networks, correlations were run between individual centrality scores of the same centrality measure across the networks. As data were not normally distributed, non-parametric correlations were used. Two sets of correlations were carried out; first, correlations were run including centrality scores of all individuals. However, researchers often draw conclusions on individual positions only within the connected component of the network, and correlations may be heavily affected by the inclusion of the centrality scores of the isolates. We therefore ran a second set of correlations including only those individuals that had a centrality score above zero. Finally, the presence and number of two types of commonly used substructures, cliques and *k*-plexes, were investigated in the weighted networks. The presence of substructures is frequently used by researchers of animal networks to identify layers in the social organisation of an animal group, and to identify individual characteristics that may underlie the formation of such layers (*e.g.* age-mates or kin may form clusters in a network). First, we determined the number of cliques, or maximally complete subgraphs, in the networks. We set the minimum size of cliques to three, as this is the smallest possible group above a dyad. Second, we searched for the number of *k*-plexes in the networks.

Table 2. Definitions of individual centrality measures (adapted from: Wasserman and Faust 1994; Newman 2007)

Centrality measures	
Degree centrality	The degree centrality is a measure of the number of social partners an individual has, and is denoted as: $$C_D(n_i) = \sum_j x_{ji}$$ Where x_{ji} is the relationship between node j and i. In directed networks both the ties coming into a node (indegree) and leaving a node (outdegree) are considered. In weighted networks, the degree centrality measures the frequency of an individual's social relationships, taking into consideration the strength rather than the number of connections.
Betweenness centrality	Betweenness centrality measures the influence an individual has over the relationships of others, and is indicated by the fraction of shortest paths that run via a particular individual: $$C_b(n_i) = \sum_{j<k} \frac{g_{jk}(n_i)}{g_{jk}}$$ Where g_{jk} is the number of shortest paths that link the two actors, and $g_{jk}(n_i)$ is the number of shortest paths linking the two actors, that contain actor i. This measure ranges between zero and the number of pairs there are in the network, not including the individual itself. This measure can be directed and weighted.
Closeness centrality	Closeness centrality is a measure of the distance between an individual and all other nodes in their network and is denoted as: $$C_D(n_i) = \left[\sum_{j=1}^{g} d(n_i, n_j)\right]^{-1}$$ Where $d(n_i, n_j)$ is the path length between node j and i. This measure can be directed and weighted.
Eigenvector centrality	The eigenvector centrality is a measure of the prominence of an individual in their network taking into account both the number of social partners and the centrality of those partners. The eigenvector centrality indicates the number of connections as a proportion of their eigenvector centrality score, and is calculated as follows: $$x_i = \frac{1}{\lambda} \sum_{j=1}^{n} x_{ij} a_j$$ Where λ is the largest eigenvector value in the network, and a_j is the eigenvector centrality of node i. This measure can be directed, but does not take into account the weights of the relationships.

In a *k*-plex, the criteria for inclusion are more relaxed compared to cliques; nodes are considered a member of a *k*-plex when they are connected to all but *k* members (Hanneman and Riddle 2005). We set *k* to 2, the minimum value possible to differentiate between a clique and a *k*-plex, and the minimum size of *k*-plexes to 4, the minimum value of *k* suggested for this measure (Hanneman and Riddle 2005). To assess the effect of filtering on the number of substructures found, we compared the number of cliques and *k*-plexes across the six networks.

RESULTS

Depending on the filtering method, the number of individuals connected to the large component of the network varied: the unfiltered network (Figure 1a), and the network with a network structure significantly different from random (Figure 1b) were the most inclusive networks. In the networks filtered both by median association strength (Figure 1c) and by mean association strength (Figure 1d), one individual was not connected to the large component, while in the network that was filtered until the giant component was close to breaking up (Figure 1e), six individuals were not connected. The largest number of isolates, nine, was found in the preferential association network (Figure 1f).

Global Metrics

The eight global network metrics calculated for each of the networks are shown in Table 3. Network metrics largely varied relative to the number of individuals connected to the network.

Table 3. Global network metrics for the unfiltered association network, and networks filtered using five different methods.

		Global network metrics							
Filtering method	Cut-off value	Binary density	Weighted density	Binary mean degree	Largest connected component	Mean geodesic	Diameter	Mean clustering coefficient	Weighted network centralisation (%)
Unfiltered	None	1.000	0.224	15.000	16	1.000	1	1.000	12
Network structure significantly different from random	0.170	0.675	0.185	10.125	16	1.325	2	0.892	16
Median association strength	0.220	0.508	0.153	7.625	15	1.438	3	0.798	20
Mean association strength	0.224	0.467	0.144	7.000	15	1.524	3	0.753	18
Giant component near breaking up	0.340	0.117	0.048	1.750	10	2.200	5	0.429	21
Preferential associations	Preferential associations	0.042	0.017	0.625	5	1.727	3	0.000	16

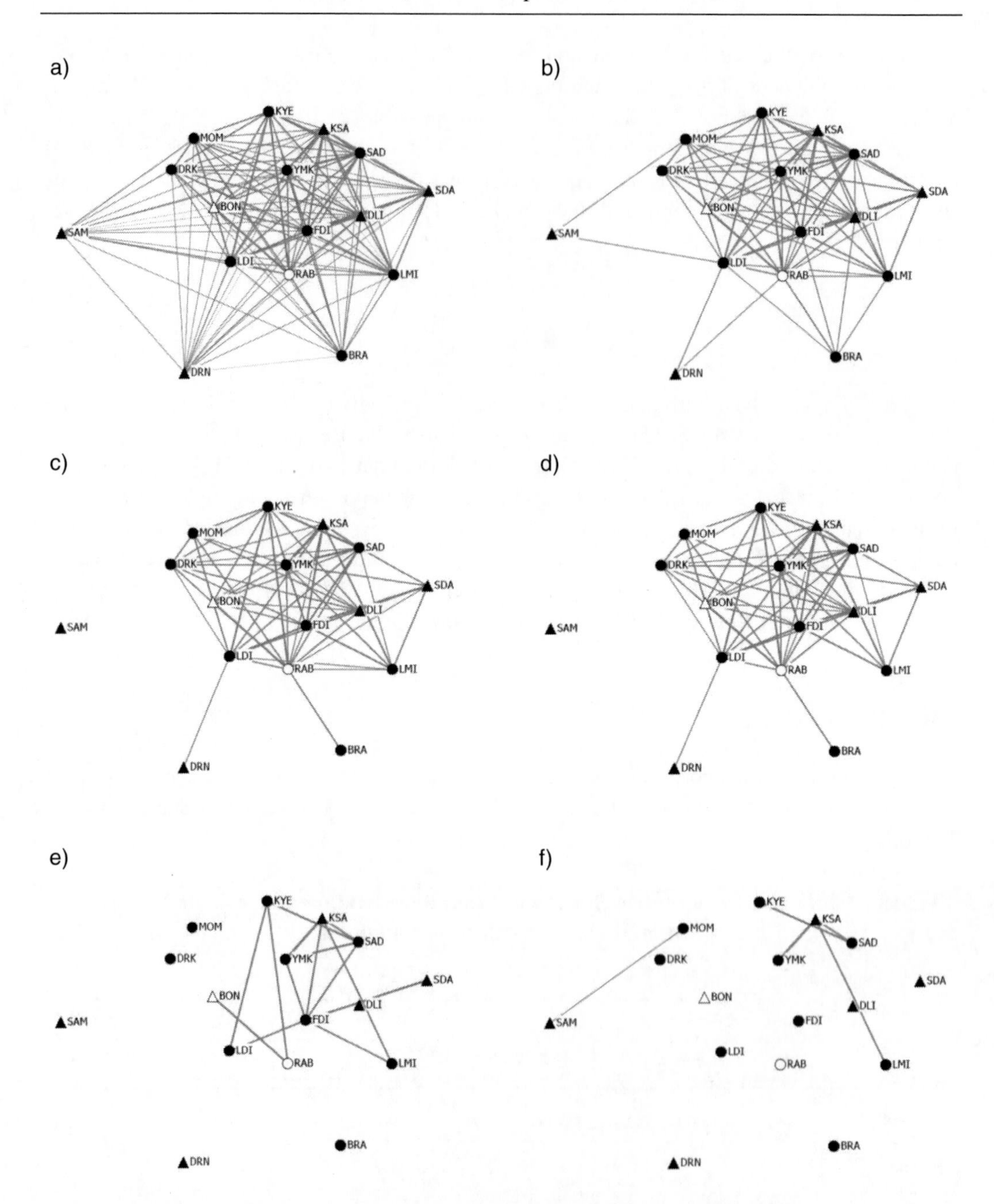

Figure 1. Baboon association networks with varying levels of filtering. Networks are: a) unfiltered; b) filtered until network structure was significantly different from random; c) filtered by median association strength; d) filtered by mean association strength; e) filtered until giant component was close to breaking up; f) filtered to include only preferential associations. Networks are laid out using spring embedding on network b. Circles represent females, triangles represent males. Black nodes represent adults, white nodes represent subadults. Edges indicate associations, with the thickness of edges indicating the strength of the association.

The diameter and geodesic increase with fewer individuals connected to the network, as would be expected, while all other metrics decrease. There were some exceptions to this general pattern: the mean geodesic and diameter of the preferential association network were relatively low. This is due to the low number of individuals that are part of the connected components. The weighted network centralisation measure also differed from this general pattern, as it varied independently from the number of connected individuals in the network.

These results indicate that global network metrics are generally affected in a predictable way by network filtering.

Individual Network Positions

The means of the four centrality measures calculated for each individual in each network are given in Table 4. On the whole, individuals became less central to their networks as the cut-off value for filtering increased. As might be expected, average degree centrality, closeness centrality and eigenvector centrality all decreased as the level of filtering increased, with fewer partners and direct relationships, whereas betweeness centrality generally increased with increasing levels of filtering, as the number of indirect relationships increases with increasing levels of filtering. On average, individuals were least central in the preferential association network.

Table 4. Average (mean ± SD) individual centrality scores for the unfiltered association network, and networks filtered using five different methods. Closeness centrality is calculated on connected components and excludes isolates. Therefore, two values are given for the preferential association network, as there are two separate components

		Individual network positions			
Filtering method	Cut-off value	Degree centrality	Closeness centrality	Betweeness centrality	Eigenvector centrality
Unfiltered	None	100.0±0.0	100.0±0.0	0.0±0.0	35.4±0.0
Network structure significantly different from random	0.170	67.5±25.2	77.9±12.8	2.2±4.1	33.4±11.6
Median association strength	0.220	50.8±25.4	41.9±4.0	2.7±5.1	32.0±15.0
Mean association strength	0.224	46.7±23.6	40.7±4.2	3.3±4.8	31.8±15.6
Giant component near breaking up	0.340	11.7±11.9	13.0±0.5	3.2±5.0	23.8±26.2
Preferential associations	Preferential associations	4.2±5.7	8.2±0.1 6.7±0.0	0.5±1.3	18.8±29.9

The results of the second analysis, in which isolates were excluded, were very similar to those presented in Table 4, in terms of the relationship between the centrality values of different networks. The only measure that appeared to be affected by the exclusion of isolates was eigenvector centrality; when isolates were excluded, the eigenvector centrality values were, on average, highest in the network that was filtered until the giant component was close to breaking up and the preferential association network, compared to all other networks. When isolates were included (see Table 4) on the other hand, eigenvector centrality values were on average lowest in these two networks. However, the results from networks without isolates are likely to be influenced by differences in the size of these networks as a result of excluding isolates. While researchers may often exclude isolates from their analyses, this may not be appropriate when making comparisons between networks, as network size is likely to affect absolute values of network metrics (Anderson et al., 1999). On the other hand, here the networks including isolates are identical in size and thus easily comparable.

Individual network positions were strongly correlated across four of the five levels of filtering (Table 5) for the different centrality measures. The unfiltered network was excluded from the correlations as this network was fully connected, and thus all individuals occupied the same network position. Only network positions in the preferential association network were not correlated with positions in any of the other networks, with the exception of eigenvector centrality scores of the network that was filtered until the giant component was close to breaking up. Overall, these results indicate that four filtering methods, *i.e.* filtering until network structure is significantly different from random, filtering by median association strength, filtering by mean association strength, and filtering until giant component was close to breaking up, would give comparable results on individual network positions, while using preferential associations would provide very different results.

In the second set of correlations only the individuals that had a centrality score above zero were included. The majority of correlations of the eigenvector scores across networks remained significant, only the correlation between the eigenvector centralities of the network filtered by median association strength and filtered until the giant component was close to breaking up was no longer significant. Out of a total of ten correlations that were run between the centrality measures of the five networks, three or four were no longer significant for the other three centrality measures (Table 5). Network positions in the network that was filtered until the giant component was close to breaking up in particular were no longer significantly associated with positions in other networks.

Taken together, these results suggest that network positions were similar if networks were filtered by median association strength, mean association strength, until the network structure was significantly different from random, or until the giant component was close to breaking up, but that the correlation of the latter may be due to the number of isolates in this network. Individual network positions in the preferential association network differed from positions in all other networks.

Despite these general similarities in centrality measures across networks, when we consider the positions of specific individuals, some differences were found, particularly between the preferential association network and all other networks. For example, taking into account the four centrality measures, RAB, a subadult female, is the most central individual in the networks filtered by the mean and median association strengths. In the network that was filtered until the structure was significantly different from random, the most central individual was LDI, an adult female, while in the preferential association network, this was

KSA, an adult male. Similarly, SAM, an adult male, was peripheral in the networks that were filtered by the mean association strengths, the median association strengths, and filtered until the structure was significantly different from random. However, in the preferential association network SAM was one of the few individuals that did have a tie. Thus, if analyses were concerned with the characteristics of central or peripheral individuals, such as age or sex, using different filtration methods may lead to very different conclusions.

Table 5. Kendall's τ coefficient for the correlation between individual centrality measures (N = 16) across the networks filtered using five different methods. As the unfiltered network was fully connected, network positions were constant across all individuals, and this network was therefore excluded from analyses. Bold values indicate a significant association. * indicates p < 0.05; ** indicates p < 0.01; * indicates p < 0.001. Shaded values indicate a correlation that was no longer significant when only the scores of individuals with a centrality score above zero were included**

	Median association	Mean association	Giant component near breaking up	Preferential associations
Degree centrality				
Structure different from random	**0.733*** **	**0.477***	**0.477***	0.011
Median association		**0.898*** **	**0.508***	0.033
Mean association			**0.527***	-0.023
Giant component near breaking up				0.386
Closeness centrality				
Structure different from random	**0.749*** **	**0.730*** **	**0.475***	0.042
Median association		**0.868*** **	**0.558** **	0.074
Mean association			**0.532** **	-0.021
Giant component near breaking up				0.353
Eigenvector centrality				
Structure different from random	**0.752*** **	**0.735*** **	**0.458***	0.138
Median association		**0.847*** **	**0.437***	0.228
Mean association			**0.454***	0.160
Giant component near breaking up				**0.600** **
Betweeness centrality				
Structure different from random	**0.462***	**0.491***	0.323	0.036
Median association		**0.831*** **	**0.507***	0.103
Mean association			**0.466***	0.208
Giant component near breaking up				0.324

Substructures

The level of sub-structuring was examined for each of the networks by determining the number of cliques and *k*-plexes in each network (Figure 2). The number of substructures found varied greatly with the level of filtering.

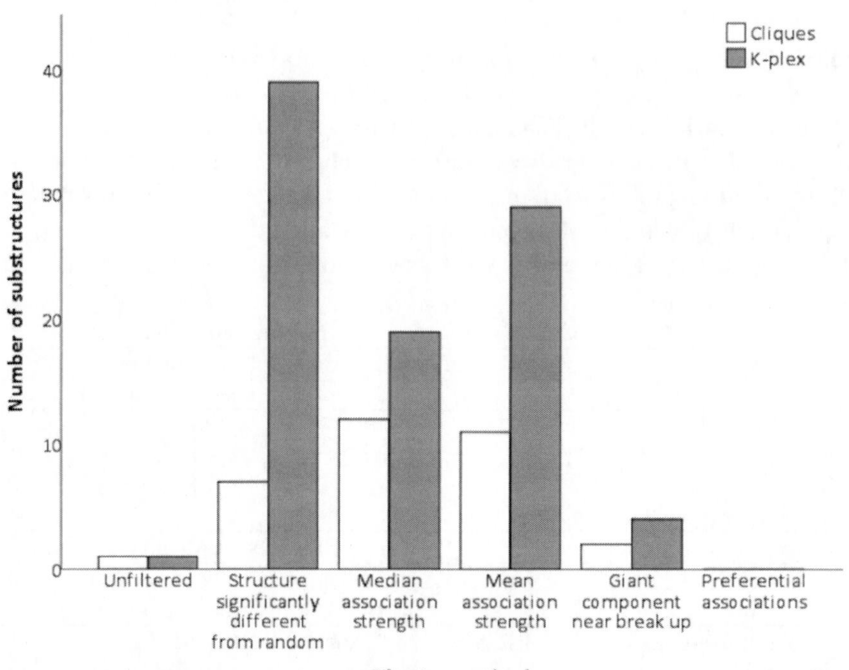

Figure 2. The number of cliques and *k*-plexes found in the unfiltered association network, and networks filtered using five different methods.

A single clique was found in the unfiltered network, and no cliques were found in the preferential association network. Most cliques were found at intermediate levels of filtering, with the number of cliques first increasing and then decreasing as the cut-off filter increased.

The variation in the number of substructures found was greater when *k*-plexes were considered. Like the number of cliques that were found, a single *k*-plex was found in the unfiltered network, while in the preferential association network no *k*-plexes were found. However, 40 *k*-plexes, the highest number of substructures, were found in the network that was filtered until its structure was significantly different from random. The number of *k*-plexes found in the network that was filtered by median association strength (19 *k*-plexes) was much lower than the number of *k*-plexes found in the network that was filtered by mean association strength (29 *k*-plexes).

These results indicate that the filtration method used can have a critical effect on the interpretation of the degree of substructuring of networks.

CONCLUSION

Our results show that while global network metrics vary with different methods of network filtering, they do so in a relatively predictable way, with metrics varying largely as a function of the level of filtering. Even the network based on preferential associations, which measure somewhat different relationships from the other filtered associations, fell into this overall pattern. In studies of animal social networks, global network metrics are frequently used to measure the impact of external factors such as food availability, or relocation, on the structure or qualities of a network, such as the cohesion of social groups (Lehmann and Boesch 2009; Dufour et al., 2011). Findings here indicate that results of such studies can be influenced by the level of filtering in a relatively predictable way: the more filtering, the less cohesive a network appears to be. This highlights the importance of using the same filtering method - or at least considering carefully the impact of using different methods - when making comparisons between network structures across time, behaviours, populations, or species.

Individual centrality measures were also found to be relatively stable across networks, as individuals had similar network positions in the majority of networks filtered by different methods, indicated by significant correlations of individual network positions across the majority of networks. In terms of network positions, it was mainly in the preferential association network that network positions greatly differed from those found in other networks. However, in studies of animal networks, researchers are often interested in comparing individuals and their particular characteristics, and how such characteristics may influence the role that they play in their network. For example, association networks of spider monkeys were found to be segregated by sex, with young males playing the role of brokers between the female and male segments of the network (Ramos-Fernández et al., 2009). Likewise, researchers have investigated how individual network position can affect the behaviour of individuals. For example, researchers found that the motivation of male Barbary macaques to carry infants is related to their position within their social networks (Henkel et al., 2010), and that the position of female marmots in their networks influences the likelihood of those females dispersing from their social group (Blumstein et al., 2009). In such studies, particularly because animal social networks are often relatively small, the network positions of individuals are important. In this chapter it was shown that while generally individuals had similar positions across networks, the identities of the most central and most peripheral individuals varied markedly depending on the filtering method that was used. These results highlight the effect that network filtering may have on conclusions drawn about the positions of individuals.

Our results show that the number of substructures that were found in a network was strongly influenced by the filtration method that was used. Both at very high levels of filtering and very low levels of filtering, no substructures were found in the network. Only at intermediate levels of filtering were substructures found. Social network analysis has frequently been used in the study of animal groups to identify layers in the social structure (Lusseau et al., 2006; Ramos-Fernández et al., 2006; Sundaresan et al., 2007; Wolf et al., 2007). Our findings show that the filtering method can have a marked impact on the number of such substructures that are found. This highlights the need for the careful consideration of the cut-off level for filtering if the main objective of the study is to investigate the substructures in a network.

Of course, some of the results presented here may be due to the particular structure of relationships in this species or population. Nevertheless, this chapter indicates the effect that different filtering methods may have on the conclusions that researchers of animal networks draw. Further development of weighted network metrics could circumvent such issues, as such measures allow for a greater differentiation between different types of relationships, making network filtering redundant. This may be especially important because is has been suggested that in stable social groups, rare interactions represent weak, but potentially important relationships (Granovetter 1973), and filtering these out may lead to misleading results (Blumstein et al., 2009). Until a greater variety of weighted network metrics has been developed, network filtering remains an essential part of many studies of animal social networks. Following the results presented here, the method of filtering should be chosen carefully in order to ensure it is appropriate to the questions asked.

Acknowledgments

Fieldwork at GGNP was made possible through the research permit of the Nigerian National Parks Service to the Gashaka Primate Project, which receives its core funding from the Chester Zoo Nigeria Biodiversity Programme. Lauren Brent gave valuable comments on a previous version of the manuscript. Funding came from University of Roehampton and The Leakey Trust. This is a GPP publication.

Reviewed by Dr. Lauren J.N Brent, Duke Institute for Brain Sciences, and Centre for Cognitive Neuroscience, Duke University.

References

Alberts, N. (2012). *Fission-fusion dynamics of olive baboons (Papio anubis) in Gashaka-Gumti National Park.* PhD thesis, University of Roehampton.

Altmann, J. (1974). Observational study of behaviour: sampling methods. *Behaviour,* 49, 227-267.

Anderson, B. S., Butts, C., & Carley, K. (1999). The interaction of size and density with graph-level indices. *Social Networks,* 21, 239-267.

Bejder, L., Fletcher, D., & Bräger, S. (1998). A method for testing association patterns of social animals. *Animal Behaviour,* 56, 719-725.

Blumstein, D. T., Wey, T. W., & Tang, K. (2009). A test of the social cohesion hypothesis: interactive female marmots remain at home. *Proceedings of the Royal Society B: Biological Sciences,* 276, 3007-3012.

Borgatti, S. P. (2002). *NetDraw: Graph Visualization Software.* 2.097.

Borgatti, S. P., Everett, M. G., & Freeman, L. C. (2002). *Ucinet for Windows: software for social network analysis.*

Brent, L. J. N. (2009). *Investigating the Causes and Consequences of Sociality in Adult Female Rhesus Macaques using a Social Network Approach*. PhD thesis, Roehampton University.

Brent, L. J. N., Lehmann, J., & Ramos-Fernandez, G. (2011). Social network analysis in the study of nonhuman primates: a historical perspective. *American Journal of Primatology,* 73, 720-730.

Cairns, S. J., & Schwager, S. J. (1987). A comparison of association indices. *Animal Behaviour,* 35, 1454-1469.

Corner, L. A. L., Pfeiffer, D. U., & Morris, R. S. (2003). Social-network analysis of *Mycobacterium bovis* transmission among captive brushtail possums (*Trichosurus vulpecula*). *Preventive Veterinary Medicine,* 59, 147-167.

Croft, D. P., James, R., & Krause, J. (2008). *Exploring Animal Social Networks*. Princeton & Oxford: Priceton University Press.

Croft, D. P., James, R., Thomas, P., Hathaway, C., Mawdsley, D., Laland, K. N., & Krause, J. (2006). Social structure and co-operative interactions in a wild population of guppies (*Poecilia reticulata*). *Behavioral Ecology and Sociobiology,* 59, 644-650.

Croft, D. P., Krause, J., & James , R. (2004). Social networks in the guppy (*Poecilia reticulata*). *Proceeding of the Royal Society London B (Suppl.),* 271, S516-S519.

Cross, P. C., Lloyd-Smith, J. O., Bowers, J. A., Hay, C. T., Hofmeyr, M., & Getz, W. M. (2004). "Integrating association data and disease dynamics in a social ungulate: bovine tuberculosis in African buffalo in the Kruger National Park." *Annales Zoologici Fennici, 2004*, pp. 879-892 41.

Dufour, V., Sueur, C., Whiten, A., & Buchanan-Smith, H. M. (2011). The impact of moving to a novel environment on social networks, activity and wellbeing in two New World primates. *American Journal of Primatology,* 73, 802-811.

Erdös, P., & Rényi, A. (1959). On random graphs, I. *Publicationes Mathematicae (Debrecen),* 6, 290-297.

Flack, J. C., Girvan, M., de Waal, F. B. M., & Krakauer, D. C. (2006). Policing stabilizes construction of social niches in primates. *Nature,* 439, 426-429.

Godfrey, S. S., Bull, C. M., James, R., & Murray, K. (2009). Network structure and parasite transmission in a group living lizard, the gidgee skink, *Egernia stokesii. Behavioral Ecology and Sociobiology,* 63, 1045-1056.

Granovetter, M. S. (1973). The strength of weak ties. *American Journal of Sociology,* 78, 1360-1380.

Hanneman, R. A., & Riddle, M. (2005). *Introduction to Social Network Methods.* Riverside: University of California (published in digital form at *http://faculty.ucr.edu/~hanneman/*).

Henkel, S., Heistermann, M., & Fischer, J. (2010). Infants as costly social tools in male Barbary macaque networks. *Animal Behaviour,* 79, 1199-1204.

Henzi, S. P., Lusseau, D., Weingrill, T., van Schaik, C. P., & Barrett, L. (2009). Cyclicity in the structure of female baboon social networks. *Behavioral Ecology and Sociobiology,* 63, 1015-1021.

Higham, J. P., Warren, Y., Adanu, J., Umaru, B. N., MacLarnon, A. M., Sommer, V., & Ross, C. (2009). Living on the edge: life-history of olive baboons at Gashaka-Gumti National Park, Nigeria. *American Journal of Primatology,* 71, 293–304.

Jacobs, A., & Petit, O. (2011). Social network modeling: a powerful tool for the study of group scale phenomena in primates. *American Journal of Primatology,* 73, 1-7.

Kasper, C., & Voelkl, B. (2009). A social network analysis of primate groups. *Primates,* 50, 343-356.

Krause, J., James, R., & Lusseau, D. (2009). Animal social networks: an introduction. *Behavioral Ecology and Sociobiology,* 63, 967–973.

Lea, A. J., Blumstein, D. T., Wey, T. W., & Martin, J. G. A. (2010). Heritable victimization and the benefits of agonistic relationships. *Proceedings of the National Academy of Sciences,* 107, 21587-21592.

Lehmann, J., & Boesch, C. (2009). Sociality of the dispersing sex: the nature of social bonds in West African female chimpanzees, *Pan troglodytes. Animal Behaviour,* 77, 377-387.

Lehmann, J., & Ross, C. (2011). Baboon (*Papio anubis*) social complexity - a network approach. *American Journal of Primatology,* 73, 775-789.

Lusseau, D. (2003). The emergent properties of a dolphin social network. *Proceedings of the Royal Society of London B (Suppl.),* 270, S186-S188.

Lusseau, D., Wilson, B., Hammond, P. S., Grellier, K., Durban, J. W., Parsons, K. M., Barton, T. R., & Thompson, P. M. (2006). Quantifying the influence of sociality on population structure in bottlenose dolphins. *Journal of Animal Ecology,* 75, 14-24.

Manly, B. F. J. (1995). A note on the analysis of species co-occurrences. *Ecology,* 76, 1109-1115.

Manno, T. G. (2008). Social networking in the Columbian ground squirrel, *Spermophilus columbianus. Animal Behaviour,* 75, 1221-1228.

Mitani, M. (1986). Voiceprint identification and its application to sociological studies of wild Japanese monkeys (*Macaca fuscata yakui*). *Primates,* 27, 397-412.

Newman, M. E. J. (2007). *"Mathematics of networks."* In S. N. Durlauf and L. E. Blume, (Ed.), The New Palgrave Encyclopedia of Economics). Basingstoke: Palgrave Macmillan.

Opsahl, T. (2009). *tnet: Software for Analysis of Weighted and Longitudinal networks.* version 2.7.

Opsahl, T., & Panzarasa, P. (2009). Clustering in weighted networks. *Social Networks,* 31, 155-163.

Ramos-Fernández, G., Boyer, D., Aureli, F., & Vick, L. (2009). Association networks in spider monkeys (*Ateles geoffroyi*). *Behavioral Ecology and Sociobiology,* 63, 999-1013.

Ramos-Fernández, G., Boyer, D., & Gómez, V. (2006). A complex social structure with fission–fusion properties can emerge from a simple foraging model. *Behavioral Ecology and Sociobiology,* 60, 536-549.

Sommer, V., & Ross, C. (2011). "Exploring and protecting West Africa's primates: the Gashaka Primate Project in context." In V. Sommer and C. Ross, (Ed.), *Primates of Gashaka: Socioecology and Conservation in Nigeria's Biodiversity Hotspot,* (pp. 1-38). New York, Dordrecht, Heidelberg, London: Springer.

Sueur, C., Jacobs, A., Amblard, F., Petit, O., & King, A. J. (2011). How can social network analysis improve the study of primate behavior? *American Journal of Primatology,* 71, 1-17.

Sueur, C., & Petit, O. (2008). Organization of group members at departure is driven by social structure in *Macaca. International Journal of Primatology,* 29, 1085-1098.

Sundaresan, S., Fischhoff, I., Dushoff, J., & Rubenstein, D. (2007). Network metrics reveal differences in social organization between two fission–fusion species, Grevy's zebra and onager. *Oecologia,* 151, 140-149.

Wasserman, S., & Faust, K. (1994). *Social Network Analysis: Methods and Applications*. Cambridge & New York: Cambridge University Press.

Whitehead, H. (1999). Testing association patterns of social animals. *Animal Behaviour,* 57, 26-29.

Whitehead, H. (2009). SOCPROG programs: analyzing animal social structures version 2.4. *Behavioral Ecology and Sociobiology,* 63, 765-778.

Whitehead, H., & Dufault, S. (1999). Techniques for analyzing vertebrate social structure using identified individuals: review and recommendations. *Advances in the Study of Behavior,* 28, 33-74.

Williams, R., & Lusseau, D. (2006). A killer whale social network is vulnerable to targeted removals. *Biology Letters,* 2, 497-500.

Wittemyer, G., Douglas-Hamilton, I., & Getz, W. M. (2005). The socioecology of elephants: analysis of the processes creating multitiered social structures. *Animal Behaviour,* 69, 1357-1371.

Wolf, J. B. W., Mawdsley, D., Trillmich, F., & James, R. (2007). Social structure in a colonial mammal: unravelling hidden structural layers and their foundations by network analysis. *Animal Behaviour,* 74, 1293-1302.

Wolf, J. B. W., & Trillmich, F. (2008). Kin in space: social viscosity in a spatially and genetically substructured network *Proceedings of the Royal Society B: Biological Sciences*, 1-7.

In: Social Networking
Editors: X. M. Tu, A. M. White and N. Lu

ISBN: 978-1-62808-529-7

Chapter 2

Hash Tags, Status Updates and Revolutions: A Comparative Analysis of Social Networking in Political Mobilization

Patience Akpan-Obong*[*] *and Mary Jane C. Parmentier
School of Letters and Sciences, Arizona State University, US

Abstract

Recent political upheavals in North Africa and the Middle East have thrust information and communication technologies (ICTs) into the forefront of global discourse. While ICT-enabled sociopolitical mobilization and activism have been notable, there is yet no conclusive evidence that the technologies significantly affect these processes particularly in developing countries with low ICT penetration levels. This chapter therefore raises three fundamental questions. First, were the mass protests in Iran, Egypt and Libya "social revolutions" as have been described in the popular media? Second, what precisely did social networking do in the Arab Spring, and can we draw conclusions about the trajectory of such events in other countries? Third, does social networking transform political mobilization in any unique way or are the constituent technologies merely another set of tools to achieve political goals in similar ways that older communication technologies did? Utilizing the Arab Spring as a springboard, we examine two unheralded mass protests that occurred in Nigeria (2012) and Bolivia (2010, 2011). These cases indicate that while social networking does indeed facilitate and galvanize political mobilization and activism, it is a catalyst rather than a significant factor in and by itself. Also, the variations in the outcomes of ICT-enabled mass protests in North Africa support this conclusion and caution a technologically deterministic emphasis on the impact of social networking on political mobilization and mass protests.

---

[*] A version of this chapter was presented at the International Studies Association Annual Convention, San Diego, April 1-3, 2012. patience.akpan@asu.edu.

INTRODUCTION

In January 2010, U.S. Secretary of State Hillary Clinton delivered what was described as a notable speech detailing U.S. policy on global Internet freedom. It came in the wake of Google's announced plan to eliminate censorship filters from its search engine in China following a cyber-attack on the Gmail accounts of Chinese human rights activists. Ma Zhaoxu, a spokesman for China's foreign ministry, defended his country's policies and denounced Clinton's comments. He insisted that the Internet in China "is open and China is the country witnessing the most active development of the Internet" (Fox News, 2010).

This exchange was indicative of a broad and complex relationship between information and communication technologies (ICTs) on one hand, and global interactions in the sphere of sociopolitical mobilization, on the other hand. The technologies, particularly the Internet (accessed from computer or mobile phones, which are dominant in Asia and Africa), have become arguably the most important global medium of communication and information exchange involving citizens, firms, governments, political parties and NGOs. In the process, the technologies engender new practices, norms and structures. The societal shift enabled by the Internet impacts sociopolitical dynamics in all sectors, requiring rigorous empirical investigation, theoretical development and methodological innovation across academic disciplines. In short, ICTs drive social change, but the dynamics vary according to the sociopolitical context, as do social and political responses from states. This is clearly evidenced in the events of 2010 and 2011 in North Africa and the Middle East that have become collectively known as the Arab Spring.

Our research is however conceptually located within a larger narrative that often begins with a rendition of the shifting dynamics of international and national politics and civic engagement since the end of the Cold War. Indeed, the end of the Cold War following the fall of the Soviet Union in the 1980s coincided with an emerging/expanding internationalism. Simultaneously, the processes of globalization deepened and expanded rapidly along with advances in developments in the technologies of transportation, information and communication (Held, et al, 1991). These technologies are, arguably, the most transformative change agents in the post-Cold War global system. They certainly have been the sources and facilitators of transformation in various facets of society and politics in varying degrees of impact around the world (Van Laer, 2007; Institute for Homeland Security Solutions, 2009). In the first place, the technologies led to a new interconnected international arrangement that departed radically from the old Westphalian statist system (Castells, 1997). The new configuration stresses the compression of space (and time) engendered by the information networks which connect people in distant locations (Akpan-Obong, 2009). It connotes a non-hierarchical network of relationships and nodes and thus frames the debates on the utility of ICTs in a new cultural politics of engagement, participation and activism. An examination of the utilization of ICTs in the broader sociopolitical context generates a deeper understanding of the extent of the impacts of the technologies on international politics and other cross-national processes. For one thing, many governments and their agencies at the national and international levels have responded to the imperatives of ICTs by adopting and integrating the technologies in their governing processes (Mayer-Schonberger and Lazer, 2007; Mossberger, Tolbert and McNeal, 2007).

The utilization of ICTs in governance is not restricted to national processes either as some countries have adopted the technologies in their foreign policies and international diplomacy (Grant, 2004). In an overview of conflict management in the network society, Wehrenfennig (2006) shows how Second Track (unofficial) diplomacy has contributed to significant success in peaceful resolution of conflicts at a greater rate than ever. While not directly crediting this to ICTs, he however notes that "social networking and social communication ... heavily impact the structural conditions for dialogue in current conflict environments" (p. 11). These technologies also feature in the discussions on peace building with many assertions about their capacities to foster better communication and therefore understanding among parties in conflict. As The United States Institute for Peace (2010) notes, mobile phones in particular, "have been used in connection with campaigns to restrain election violence, reduce corruption, develop the news media, and support counter-insurgency to name just a few" (p.1). It suggests that though overall outcomes have been mixed many successes have been recorded. Given the supposed egalitarian characteristics of ICTs, it is inevitable that their impacts would be more evident in the sociopolitical arena as civil society organizations and citizens utilize the technologies for mass mobilization, social-political activism and civic engagement. This is the context in which much of the popular media and emerging scholarly research from the multidisciplinary intersections of international relations, ICTs, communication and democratization have cast the Arab Spring as social revolutions. Not only were the 2010 and 2011 mass protests in North Africa and the Middle East tagged as revolutions, their processes and outcomes were also attributed to the power of social networking sites such as Facebook and Twitter. However, it is important that we interrogate the very notion of the Arab Spring as a social revolution. And therefore, drawing from the works of Theda Skocpol (1979) and Ted Gurr (1970), we examine the degree to which those events could be considered social revolutions or social movements in the classic definition of these concepts. We also seek to examine the role that social networking played in the mass protests and the potential for this medium of communication to be utilized by other groups as veritable tools for political mobilization and civic engagement. In the process, we redirect fixation on the Arab Spring to equally significant but unheralded mass protests in other parts of the world where social networking could have been critical to galvanizing popular resistance movements. Specifically, we examine the Occupy Nigeria protests against the removal of subsidies on gasoline that occurred in January 2012, street protests in Bolivia over a similar government policy, as well as the more recent marches in the Amazon against the Bolivian government.

We begin with a general discussion of social revolutions and social movements to facilitate our understanding of whether we can refer to the Arab Spring as revolutionary and also if the utilization of social networking was a driving factor of mobilization and protests. In the next section, we review the literature on social networking and political mobilization. This is followed by the case studies on Nigeria and Bolivia. We conclude with a discussion and analysis of the role of social networking in political mobilization especially in countries where ICTs do not constitute apparently critical factors in the mobilization, or are sparsely diffused.

Social Revolution and Social Movements in Historical Context

There are specific distinctions between social revolutions and social movements. Social movements, sometimes referred to as civil society organizations, occur within "civil society, the area of public life that involves groups of people in activities outside the formal arena of politics" (Massey, 2012: 182). Groups agitating for change or resistance in various aspects of sociopolitical structure engage in contentious politics in the sphere of civil society. These groups constitute social movements which Snow & Soule (2010) define as "collectivities acting with some degree of organization and continuity, partly outside institutional and organizational channels, for the purpose of challenging extant systems of authority, or resisting change in such systems" (p.6). Defining characteristics of social movements include: a challenge to authority, pursuit of collective goods and presence of perceived grievances. Social movements often involve a coalition of varying groups within society. They are likely to emerge in response to interruptions in daily activities and everyday life events in a way that make people feel disentitled (Snow, et al, 1998). This implies that the event that triggers the emergence of social movements is sudden but the groundswell that leads to mass protests may be gradual. Gurr (1970) explains that social movements and revolutions ensue when there is a disparity between people's expectations and reality in any given area. This could occur when expected values of freedom and equality are not experienced by different social groups such as women and minorities. Authoritarian and democratic regimes react differently to social movements. One is likely to suppress and quell at all costs, including loss of lives, while the other is more likely to create laws that protect the right of the people to protest.

Success of a social movement is often defined in terms of its character particularly in "how well it frames and articulates issues and grievances; how well its leaders serve the organization; how well resources are gathered and used on behalf of the movement; and how well it gains acceptance and supportive public opinion" (Massey, 2012: 183). However, often missing from this definition is the question of causality as it is difficult to attribute a specific change to the activities of social movements (Massey, 2012). Even when the observed change aligns with the expressed goals of social movements, it is still difficult to find causality especially in a "complex society with many institutions vying to exert power" (p. 183).

Social movements often fail to achieve their stated goals for many reasons such as internal organizational conflicts, a stronger opposition or the eruption of events that render irrelevant the causes for which a particular social movement was advocating. Regardless of failure or success in outcomes, social movements already succeed by conveying issues to the public arena thus generating debates and/or attention and setting the agenda for discourse. Also, social movements may have intended or unintended effects "but the course of social change is different from what it would have been had people not pursued their cause. It may generate counter-movements, be opposed and suppressed by authorities, be discredited and undermined, be forced to shift tactics, or accept an outcome less than what its participants had hoped for. This too, matters" (Massey, 2012: 183). For social movements then, even if expressed goals are not achieved, their existence and activism are considered success already (Massey, 2012).

Social revolutions are similar to social movements in that they alter the course of social change, but perhaps that is where the similarity ends. Social revolutions have more fundamental impact on the sociopolitical structure. Indeed, this is the defining characteristic of social revolutions, as argued by Skocpol (1979) in her examination of three social revolutions (French, Russian and Chinese). Social revolutions transform "state organizations, class structures and dominant ideologies" and can also have international impact especially if the state in which they occur is "large and geopolitically important" (Skocpol 1979: 3). For an upheaval to be described as social revolution, it must be rapid, sudden and lead to actual change in the country's social and political structure. Tilly (1984) and Johnson (2010) contribute to the definition of social revolutions by integrating the elements of political violence, collective action and change. As Johnson (2010) notes, social revolutions occur when an "existing social system comes into crisis." Violence becomes an integral process of social revolutions and the outcome is a change in a society's "value-orientation."

Accordingly, social revolutions have "distinctive pattern of sociopolitical change" (Skocpol, 1979:4) and must also be class-based, emerging from below. In this sense, social revolutions are different from other forms of political upheavals such as riots, mass protests or coups. For one thing, by definition, social revolutions must succeed – achieve their clearly defined goals, often of upending the existing political or social systems. This is a major distinction from social movements which are not evaluated on success/failure criteria. Rather, social movements are considered successful even when they fail to achieve their stated goals because, regardless of the outcome of their activism and advocacy, they have already succeeded by conveying their issues and grievances to the public forum. On the other hand, social revolutions must succeed in disrupting the sociopolitical and structural status quo not only within the specific geographical context in which they occur but must trigger similar events in other countries. Indeed, while we still need to determine the degree to which the 2011 mass protests in Egypt can be considered a revolution, the general argument is that they were activated by the events in Tunisia only a month earlier.

Despite the mass protests in Iran in 2009, the Arab Spring, arguably started in Tunisia (Noor, 2011) when a 26-year-old street vendor, Mohamed Bouazizi, set himself on fire to protest the seizing of his wheelbarrow by local authorizes. If any dimension of the Arab Spring can be considered revolutionary, perhaps the events in Tunisia approximate Skocpol's definition even though there was no massive violence. Besides the resignation of the national leadership, the structure remained unaltered in any fundamental manner, but that singular event signaled to aggrieved groups in other countries in the region about the power of the people to protest an existing regime out of office.

While social movements are characterized by coordinated organization and participation by a broad spectrum of society and interest groups, a social revolution is distinctive in the sense that it is violent, involves conflict between specific and self-identified classes, and must lead to fundamental structural change in the sociopolitical constitution of the society. By these definitions, it is clear that the events in Tunisia, Egypt and Libya, though they had elements of violence and some conflicts between the political leadership and the people, cannot be described as social revolutions. In the first place, while the mass protests (and a near civil war in the case of Libya) led to change in political leadership, the structures of inequality and suppression against which the people protested remain intact. In Egypt, the military leadership that stepped in to replace Mr. Mubarak and his administration remains in power more than two years since thousands of Egyptians laid siege to Tahrir Square in Cairo. Recent reports out of Egypt continue to give meaning to the expression: "the more things change, the more they remain the same." As one commentator at a 2012 conference in Atlanta

Georgia put it, the "Arab Spring is over, but Egyptians are still waiting for the summer." Thus these events fit the criteria more closely associated with social movements, since they emanated from a broad spectrum of society and challenged the political status quo even if there was no structural change.

A detailed study of the mass protests in Tunisia, Egypt and Libya is certainly outside the scope of this chapter. Our focus, instead, is on the manner in which the events in these countries in 2011 renewed research interest in the role of social media in political mobilization. Of course, suppositions about the implications of the Internet are traceable to the 1990s though earlier research was mainly on the technology's socioeconomic impact and how developing countries could harness it to achieve macroeconomic goals (Akpan-Obong, 2009). The expansion of Internet access in the 1990s and emerging social networking sites such as MySpace, Facebook, Twitter and Blogspot generated additional interest in the sociopolitical implications of the technologies and applications for civic engagement and participation. At the same time, there were studies on control and usage of the technology by governments to repress citizens' rights. While the technologies empower citizens, they have also been used as tools by those in power (as was the case in Iran in 2009). Obviously it is too early to reach any conclusions about the ultimate consequence of social networking in the political sphere. We argue however that there is enough research in the field to allow for tentative comparative studies aimed at isolating possible variables to determine the degree to which social networks are critical to political transformation in a given context. While the Arab Spring and the role of social networking continue to generate research interest and excitement, we take a different track by considering Nigeria and Bolivia, two countries that experienced relatively significant but unheralded mass protests in the wake of the Arab Spring. We seek to understand if social networks made any significant contribution to those protests. This is premised on the assumption that if social media are significant catalysts or facilitators in some countries, they are likely to exhibit the same effects in others. If they do not, then that fact may highlight more fundamental factors that explain the variations in outcomes. Before we get into the case studies, we set the context for the analysis that follows by presenting a brief review of recent literature on social networking and political mobilization.

SOCIAL MEDIA AND POLITICAL ACTIVISM

The case has been made that the Internet is not inherently liberating, and that there is significant evidence to support this claim. Deibert & Rohozinski (2010) argue that the Internet is inhabited by a myriad of actors, and is regulated by many, often conflicting behaviors. These actors have diverse agendas, and the results of their participation are not necessarily more freedom of expression and political action. Actors typically range from governments who actively seek to suppress political opposition, to groups who interact in intimidating and suppressive ways. Golkar (2011) has noted the Iranian government's enhanced ability to identify activists against the regime and shut down social media. At the same time, social media, particularly through texting, played a significant role in the so-called Green Revolution and protests following the Iranian elections in 2009. Golkar (2011) characterizes the Internet as Janus, with two faces: one of government control and the other a vehicle for opposition and expression – two "inseparables."

It is as a vehicle for opposition by civil society and members of oppressed groups that most studies on the Internet have been focused. Acknowledging the benign use of the technology as well as its functionality as a tool of repressive regimes, Diamond (2010) describes it as a "liberation technology," with examples of expanding online political activism and censorship in countries around the world. Howard and Hassain (2011) have argued that in North Africa, activism on social networks served as a major catalyst, even bringing down regimes despite the absence of a centralized leadership among activists, changing notions of how social and political movements coalesce. Howard had made the case in an earlier work (2010) that in the Muslim world the Internet and digital media offer a whole new arena for protestation and political discourse, including transnational communication and identity. This follows previous research that predicted significant outcomes for online political activism. Dartnell (2006) depicted the power of images in political action, and characterized the Internet as a new "media scape" for the development of political images and messages. The transnational space available through the Internet was also highlighted as a significant factor in empowering oppositional groups. Diasporas have created online communities, and in some cases contributed to homeland insurgencies and political movements, greatly expanding this transnational political space (Brinkerhoff, 2009; Odutola, 2012). This idea of an expanding space for political activity was demonstrated by Murphy (2006) to be significant in political change in the Gulf Arab states, though at that time significant control by the states limited that space, as noted by Deibert & Rohozinksi (2011). Even with government control and suppression, the blogosphere in Iran has been depicted as a "space for contention" between society and the state, significant despite the crackdown on the Green Revolution in 2009 (Serberny & Khiabany, 2010). This corresponds with Howard's idea of the Internet as providing "ideational space" for political and social thought (Howard, 2010).

From these perspectives, it is clear that social networking constitutes part of a broader phenomenon of sociopolitical processes on the Internet. These include political websites, blogs, new digital media, e-government, and social networking sites such as My Space, Facebook, Twitter, Yahoo! Groups and Google Groups. It is important to note these elements of the online space do not exist in isolation: blogs can be an interactive part of a website, Facebook connects with blogs and websites, the totality of which aggregates in the *networking* aspect of the activities occurring in cyber space. Undoubtedly, it would be restrictive to isolate the study of any one of these elements in a particular country, as this might exclude the most salient factors in the role of digital networks in that context. For instance, blogs and websites link to more clearly defined social network sites such as Facebook and Twitter which also loop back to different spaces on the Internet such that one knows the point of entry but not always where the journey (browsing, surfing or trolling) will terminate. It would therefore be futile to study the effect of one media source – for instance, Facebook – in isolation of other media forms and processes on the Internet. Indeed, as Mudhai et al (2009) note, the study of digital media in Africa and their varied effects on democratization in the region, requires, as a starting point, a critical understanding of not just the interactions of other new media but also the role of older media. This is underscored by the fact that social networking technologies are not widely diffused in African countries, though great strides have been recorded in recent years (ITU, 2012; Internet World Stats, 2012). Thus, the imperative to consider the unique context of each country is an insight that much of the literature emphasizes. Serberny & Khiabany (2010) accentuates this point by arguing that to understand the significance of blogging in Iran one must understand the political context and political mobilization historically and offline. Also, Lim (2012) has shown that the so-called Facebook or Twitter Revolution in Egypt was anchored on a long history of political organization and opposition to the state. Thus, while social networks played an interesting and perhaps crucial role in the uprising that overthrew Mubarak, it could not have happened without the offline organization and activism that preceded it by many years. Goggin & McLellan (2008) argue that many previous studies focused on Western democracies and ways through which established political systems utilized and were influenced by the Internet. They suggest that case studies from many different societies are vital to an understanding of the different ways in which the Internet is utilized socially and politically.

While the need to study the phenomena in each unique political, socioeconomic and cultural context is recognized and indispensable, the question is: how do we get beyond single case studies to understand some common dynamics and transnational or regional aspects to social networks and political mobilization? Howard (2010) attempted to address this question in his impressive work on information technologies and politics in the Muslim world. His scope was broad, including the entire Muslim population, covering everything from e-government to Muslim blogging and identity. He found that, over all, the Internet was becoming a critical space for political discourse, idea formation, and both governmental control and contestation from society. Garrett (2012) has noted that this research is being conducted by multiple disciplines, making it difficult to assess the state of knowledge. He suggests a framework using social movement literature that offers the categories of "mobilizing structures", "opportunity structures", and "framing processes" (Garrett 2012: 203). Based on this framework, along with a review of some of the older research from the late 1990s and early 21st century, he has placed the findings in a theoretical social movement framework that includes mobilizing structure, opportunity structure and framing processes. Thus, ICTs offer a *mobilizing structure* by lowering the costs of activism, and the promotion of collective identity and community. *Opportunity structure* describes the unique attributes of the technologies and the difficulty that states have in regulating them. Finally, the *framing processes* refer to the ways in which discourses about social movements or causes develop. These framing processes are similar to Howard's notion of idea formation (Howard 2010). Garrett further notes: "Used in different contexts, technologies yield different effects" (Garrett 2012). We suggest also that technologies are used in different ways, not just in different contexts. It is therefore important to examine which aspects of social networking are utilized, and how they relate to each other, and other aspects of the Internet as an approach to understanding the impacts of these technologies on political mobilization and social protests.

The literature so far has shown that ICTs and their facilitating of social networks can both engender and inhibit political mobilization depending on each political, social and economic context and timing. While case studies continue to be the main method of understanding the phenomenon, we agree with Howard (2010) that broader studies and comparisons will be useful in understanding the potential variable of online social media in political change. Furthermore, we argue that as important as it is to understand the significant variable in political change, such as unfolded in the Middle East in 2011 and 2012, it is equally important to study and compare cases where there is apparent lack of influence of social networking on political activism. To investigate the utility of this proposition further, we examine some events in Nigeria and Bolivia, two countries that share certain attributes. These commonalities include a reliance on natural resources with global demand, an elite power structure with socioeconomic inequities, and ethnic divisions. Also, the countries experienced mass protests on similarly aligned issues, and social networking was utilized in varying degrees during the protests. But Bolivia and Nigeria are also different in terms of the nature of these ethnic divisions and their influence on the political system, as well as the ways in which the countries have developed. However, together, they present a contrast to the Arab Spring in the sense that while Nigerians and Bolivians had distinct grievances, took to the streets and utilized social networking tools to varying degrees, their protests were more or less unheralded and ended without any structural change in political leadership (besides some modest policy modifications). Nigeria and Bolivia therefore present an interesting juxtaposition that highlights what may very well be an exaggeration of the power of social networking in political mobilization. Alternatively, the contrast may underscore a stronger effect of other factors on political mobilization and mass protests particularly in contexts where social networking is sparsely diffused.

CASE STUDIES: NIGERIA AND BOLIVIA

#OccupyNigeria and Fuel Protests, 2012

On Friday, March 2, 2012, Chief Chukwuemeka Odumegu-Ojukwu was laid to rest in his home town of Nnewi in eastern Nigeria. He spent much of his last 40 years as a civilian, but he was given a full military burial because he was a soldier, first as a colonel in the Nigerian Army. He also became a general in the Biafran military, an institution he created when he led much of Eastern Nigeria in a campaign of self-actualization. This resulted in a three-year bloody civil war as the rest of Nigeria, led by General Yakubu Gowon, resisted the secession bid (Uwanaka, 1981; Ori, 2009). Though the factors that led to the Nigerian Civil War (1967-1970) remain unresolved (Mansour, 1993; Uwechue, 2004; Ojeleye, 2010), it is unlikely that Nigerians will ever again go to that extreme to address sociopolitical or economic conflicts. This is not for lack of strong forces that would potentially generate a similar violent outcome in the country. For instance, in June 1993, the military government of General Ibrahim Babangida annulled the election of Bashorun M. K. O. Abiola, the presumed winner of a presidential election widely adjudged as fair and free, probably the first of such in the country (Omoruyi, 1999; Falola & Heaton, 2008). Abiola was from the Yoruba ethnic group located in the south-western part of the country. Southerners, but mostly Yorubas, protested the annulment. There were all the signs of a protest, Nigerian style: placards, burnt properties (buildings and vehicles) and labor strikes by trade unions (who were pro-Abiola) and the economy ground to a standstill. However, these occurred mainly in Lagos, the economic hub of the country, the most populated mega polis and a bastion of the Yorubas. Weeks later, the protests fizzled and most people returned to business as usual.

On January 1, 2012, Nigerians woke up to the news that President Goodluck Johnathan had, despite his promises to the contrary, removed subsidies on the consumer price of gasoline in the country (CNN Newswire Staff, 2012; Flood, 2012). This raised the cost of the product from N65 (US$0.40) to N141 (US$0.87) per liter, a major jump in a country where

70% of the population live below two dollars a day. Many Nigerians were upset and set out to stage what became known as #Occupy Nigeria in the model of the Occupy Wall Street Movement in the United States, an effort that arguably drew inspiration from the Arab Spring. Organizers of #OccupyNigeria also shaped their modus operandi after the 2011 “revolutions” in Tunisia, Egypt and Libya that led to the fall of Zine el-Abidine Ben Ali, Hosni Mubarak and Muammar Gaddafi, respectively (Olofinlua, 2012). A main feature of those revolutions was the utilization of social networks on the Internet thus earning the tag of Revolution 2.0 (in reference to the interactive nature of social media and the departure from earlier modes of social and political protests). It has also been described as “Facebook Revolution” (Ghonim, 2012; Herrera, 2012).

#OccupyNigeria activists therefore massively used social media, mostly Facebook and Twitter, during the six-day nationwide protests and shut-down and many Nigerians credit social media for the execution and outcome of the protests (Olofinlua, 2012). Government officials such as the president and his ministers, specifically the Minister of Finance and former vice president of the World Bank, Ms. Ngozi Iweala-Okonji, used social media extensively to push to the public consciousness the government rationalization of the decision to end the subsidization of gasoline price. On the other side of the arguments were Nigerians who besieged President Jonathan’s Facebook page with persuasive, critical, angry and generally negative posts on his “Wall.” The conversation was mostly one-sided as the President did not respond to the hundreds of comments from his “likers” (numbering in the thousands). While the technology allowed for interactivity (in the mold of Revolution 2.0) the dynamics of communication between the Nigerian governing class and the rest of the country remained traditionally unidirectional. Thus a reversal in the direction of information flow as enabled by the technologies did not enhance the responsiveness of policymakers. The number of photos and other visual images from the six-day strike posted on Facebook was significant only for their shock value. A most tellingly gruesome of these images was a series of photos that depicted a fatal confrontation between armed military personnel and a protester. The series began with the first photo showing the man being stopped by the military men, all the way to the last shot with the man’s fatally wounded and mutilated body. Many similar photos of brutalities captured by smartphones were regularly posted to Facebook by protesters. Other images showed a group of fashionably dressed young women armed with cameras and smartphones on one side of the road. On the other side of the road and across a barricade stood a group of heavily armed soldiers watching the crowd seemingly daring them to approach. A similar photo in contrast was that of a group of Muslim women (so identified by their hijab head covering) sitting in the Islamic prayer posture. They held up a sign that read: “NIGERIA IS NOT/ANIMAL FARM/it is our right /To protest.”

While there were these and many other digital activities and online activism in the #OccupyNigeria movement, in no way did they mirror Tunisia, Egypt or Libya. First, while images of the protests were uploaded to social network sites, and both activists and government officials used the networks to get their views across, the effect was minimal in Nigeria in terms of transferring to offline activities besides sporadic street demonstrations and the defiance of a state-imposed curfew in Kano, a northern city. Even in these examples, there was no confirmation of any assumption that demonstrators were responding to online activities beyond the fact that the technologies did enable communication between groups of protesters. Also, notices about #OccupyNigeria events in Nigeria and in some UK and US cities were posted to Facebook. Nonetheless, there is no evidence that people took to the streets in their different cities as a result of these online messages and announcements. The social networking sites provided access to information about the events but any conclusion that they galvanized concrete offline actions can only be speculative. Second, much of the negotiations by the Nigerian protesters and government officials continued to occur on the ground mostly between trade union leaders (ostensibly on behalf of Nigerians) and federal government representatives in Abuja, the federal capital.

Third, the movement dissipated after labor and government reached a compromise on the cost of gasoline even though the protesters did not achieve their goal of getting the Jonathan Administration to revert to the pre-Jan. 1, 2012 price of N65 per liter. Indeed, #OccupyNigeria had quickly become a larger protest than simply the price of gasoline. The removal of the subsidy activated other grievances particularly concerning the magnitude of corruption by Nigerian politicians. "While it started off as a protest against the increase in fuel prices, #Occupy Nigeria was really about Nigerians saying they would no longer stand for the government's corruption" (Nesbitt-Ahmed, 2012). Other issues that came to the surface were the general state of insecurity in the country (especially as heightened by killings and bombings by an Islamist militant group, Boko Haram), high unemployment rate, poor road conditions, inadequate access to healthcare and the deplorable state of education in the country. These issues were hardly addressed during the negotiations. Faced with threats of job loss, Nigerians continued to grumble but quietly went back to work. They made a slight gain as the new price of gasoline was reduced to N97.00 (from the proposed N141.00) but other grievances went unresolved. However, by the time the nationwide strike was called off by leaders in the trade movement, the Nigerian economy had lost about $3.1 billion (Nesbitt-Ahmed, 2012) due to the shutdown in economic activities (banks, shops, manufacturing firms, hospitals and gas stations were closed during the six-day strike). Incidentally, part of the issues that the #Occupy Nigeria protesters were concerned about, the threat of Boko Haram, had bloomed into a nationwide insecurity epidemic with several incidents of bombing, suicide attacks and massacres in churches in northern Nigeria. By the end of 2012, Boko Haram had become on one hand, a coordinated terrorist group, and on the other a rag tag army of splinter groups and diverse agendas and targets. A fatal attack on the convoy of the Emir of Kano, Alhaji Ado Bayero, in January 2013 took the threat to new levels of insecurity transcending religion and ethnicity. A local government chairman and three palace guards were killed during the attack believed to have been an assassination attempt on the Emir (Ameh, 2013). An Emir is a religious, cultural and political position among Muslims particularly in Northern Nigeria. A protest against Christianity and Western civilization, as the name of Boko Haram connotes, would consider an attack on an Emir extremely sacrilegious. But then, some of the Boko Haram factions "privilege a version of Islam that regards as transgressors those who do not abide strictly to the teachings of Allah. The net effect is that faithful Muslims are often targets of its wrath" (Maiangwa & Uzodike, 2012: 3). Of interest to this discussion is the level of social networking technologies that Boko Haram groups have used in communicating their threats and messages. It is common for these groups to post video clips of their attacks on YouTube as well as well as use the various social networks to communicate their messages to each other and the public, namely the federal government. Indeed, it has been suggested that "the growing use of the new media (the Internet and mobile phone) is rapidly contributing to the success of the group's violent agenda" (Musa, 2012: 111). In this process, social networks become truly the "technology of good and evil" (Akpan-Obong, 2011) providing unlimited capacity to groups on both sides of the social and political divides in the Nigerian society. This point may then affirm one of the assumptions made in this chapter: social networking may reduce the transaction cost of protests and other forms of political mobilizations but they do not by themselves set the agenda. This is also the case in Bolivia where there is an absence of a significant connection between political activism and social networking in ways that achieve outcomes fundamentally different from previous eras of activism in the South American country.

Bolivia: Indigenous Politics and Protests, 2011 – 2012

In 2005 Bolivia made history by electing the first indigenous leader into office in a Latin American country where the majority identifies as indigenous (Cameron & Hershberg, 2010). Evo Morales, an Aymara Indian with peasant roots and former leader of the coca growers association, rose to power supported by MAS (Movimiento al Socialism, or Movement to Socialism; also meaning "more" in Spanish). The party was unique in that it represented a wide coalition of groups, not just indigenous interests (Cameron & Hershberg, 2010). The ability to coalesce so many different groups in the movement had much to do with the perceived inadequacy of the previous government to address Bolivia's socioeconomic inequities (among the highest in a region known for severe gaps in income), and recent policies aimed at privatizing Bolivian resources such as water and natural gas. Thus, while Morales had broad support from indigenous groups, there were crosscutting issues as well that propelled MAS onto the national stage. However, by 2012, there were several notable protests against the Morales government, causing a fracturing within MAS, while also pitting powerful indigenous groups against the state. These protests have been manifested by large numbers of people in the streets and some violence, and neither appears to have been facilitated significantly by online social media and networking.

The first significant popular protest occurred in December 2010 when Morales announced that the government would cut fuel subsidies to save money, a policy that would have raised fuel prices by 73% (Romero, 2011). Thousands of people in Bolivia's main cities protested in the streets, throwing rocks at government buildings, damaging public property, blocking highways, and public transit workers went on strike. Five days later the government dropped the proposed policy. Interestingly, this is a similar dynamic to the protests that led to the creation of MAS and the ousting of the pre-Morales government – protests over government policies perceived as adversely affecting the poor and marginalized, which was the foundation for MAS and Morales' rise. The protests in the early part of the 21st century disrupted spontaneously and through "on the ground" social, political and labor networks, as they did more recently. While online social networking was not yet extensive years ago, it was certainly available in 2010. (In 2012, nearly 20% of Bolivians were Facebook subscribers, exceeding Nigerians, at less than four percent – See Table 1 below.) Bolivia has a history of social movements playing out in the streets and organized at the local level. Online social networks do not seem to have such resonance or pull, at least not yet (Road rage, 2011). There is slightly more discussion of the proposal to build a transnational highway through Bolivia on Facebook, Twitter and in the Bolivian blogosphere. However those who are discussing are not always the same as those marching on the streets, such as the thousands of Amazonian Indians from several tribes who massed in the summer of 2011 for a two-month trek from the Amazon Basin up to La Paz, the country's capital, to protest the government plan to build a highway through their lands.

The Villa Tunari – San Ignacio de Moxos Highway, expected to connect Brazil to the Pacific Ocean and create a new infrastructure for Bolivia, was to traverse several hundred miles through the middle of the Isiboro Secure National Park and Indian Territories (TIPNIS), cutting down rainforest in its path (Road rage, 2011). In response to the two-month trek to the capital, Morales cancelled plans for the road and outlawed highway construction in the park. Then, in January 2012 another march on the capital occurred, from a different indigenous organization representing tribes from the TIPNIS region. They were demanding that the road should, in fact be built, citing economic development benefits (Kenner, 2012). Interesting Bolivian political scenario, for sure, but the point is that these protest marches were organized in the Amazon at the local level, where ICT penetration is low, via traditional methods of social and political mobilization. In both cases – the fuel subsidy and the road project – the protesters achieved their goals without making the cause a "Facebook Revolution." Of course, there is no evidence so far that social networking did not play any role at all. Indeed, the fact that there is not much information about the utilization of these technologies in the three Bolivian protests that we have cited may be a good indication of the insignificant role that the technologies play in political mobilization and mass protests. This is not, however, to say that the technologies do not contribute at all; research is yet to confirm an unequivocal link between the two processes and the degree of impact.

In the Bolivian blogosphere, there is only occasional commentary on the indigenous road protests, and much more on other controversies such as the role of coca in the Bolivian economy, and criticism of Morales and his increasing assumption of power through constitutional change (see, for instance VozBoliviana). On Facebook there is Bolivia Politica, which seems to contradict the government, and has recent several mentions of the TIPNIS conflict, in which the members reporting are opposed to the highway. They are not, however, leaders of the indigenous movement against the construction. This group and several others are also on Twitter, sending messages about a myriad of issues but not particularly about the Amazonian conflict. It has been noted by one observer that politicians in the Bolivian government are increasing their presence on Twitter (Rivera, 2011). Thus, it appears that in Bolivia the political elite are utilizing social networking, but the mass movements against specific policies are occurring physically in the streets, mobilizing without the facilitation of social networking sites.

This again underscores some of the literature which insists that historical, political and social contexts matter, and that although the leaders of the indigenous movements have access to the Internet, as evidenced by their websites, there is no proof that social networking is enhancing or replacing local grassroots mobilization. The recent protests follow patterns and manifestations similar to pre-social networking days in the country, despite the fact that in Nigeria, there was also considerable activity on social network sites. The results were similar, with the governments responding to quell the protests, both through negotiation and military force. It is also useful to note that state response is critical to the outcome of all mass protests with or without social networking technologies. In Tunisia, the ruling class quietly resigned and exited the country and Tunisian mass protests quickly subsided. In Libya, the central government dug in its heels, as it were, thus leading to a protracted and violent outcome that involved military intervention by the international community. While this is not the focus of our article, we suggest in passing that perhaps the response of the state whether in Iran,

Egypt, Libya, Nigeria or Bolivia had a stronger impact on the outcome of the social political upheavals than social networking did.

DISCUSSION AND ANALYSIS

The case studies (Nigeria and Bolivia) underscore our tentative argument that while social networking facilitates, to some degree, the protests and resistance to policy change, activism and the relative response by the government occur on the ground and in more conventional manner. Also, though the debates on how to define or characterize the Arab Spring are still unfolding, it is clear that the events in Tunisia, Egypt and Libya did not constitute social revolutions in the classic sense nor did social networking achieve significant outcome independent of other factors. Not in contention however is the recognition that the technologies facilitated the articulation of the mass protests, the ability to attract the sympathy and attention of the international community and reduced the cost of participation. We argue however that the integration of social networking in these protests was not radically different from the manner in which technologies have historically contributed to other sociopolitical upheavals. A commonly cited example is the printing press and its catalytic role in the Reformation in 15th and 16th century Europe. Indeed, an examination of the level of diffusion of the technologies that enable social networking such as Facebook – cell phones and Internet – in Tunisia, Egypt, Libya, Nigeria and Bolivia preclude a hasty conclusion that social networking had a significant impact.

Table 1. Level of social networking and literacy rate, 2012

Country	Facebook subscribers (% of population)	Mobile cellular subscription per 100 inhabitants*	Internet usage (% of users)	Literacy rate(% of population)**
Tunisia	28.9	116.9	36.3	88.9
Egypt	14.5	101.1	35.6	72
Libya	10.0	155.7	11.3	97.7
Nigeria	3.4	58.58	28.4	66.6
Bolivia	19.6	82.8	15.9	90.7

*The data are for 2011.

**The data are for 2010.

Sources: Internet World Stats: Usage and Population Statistics, 2013; International Telecommunications Union, ITU (2013); United Nations Development Program, UNDP (2013); Nigerian Communication Commission, NCC (2013). Millennium Development Goals Database, United Nations (2013).

As Table 1 shows, while Libya had the highest cell phone diffusion in 2011, at 155.7 for every 100 inhabitants, far exceeding Egypt's 101.1, the percentage of Libyans on Facebook was just slightly more than 11 percent. Tunisia recorded 116.9 cell phone subscriptions for every 100 inhabitants but Tunisians were more numerous on Facebook (28.9%) and on the Internet (36.3%) in 2011 as well as in December 2010 when their mass protests began (Nesbitt-Ahmed, 2012). It is also significant that besides Libya, Tunisia had the highest literacy rate among the other three African countries on the table. Could the combination of high cell phone diffusion, Internet penetration and Facebook subscription account for the quick and relatively non-violent resolution of the civil unrest in Tunisia? As attractive as this explanation might be, the trajectory of events in the two Arab countries and in our "contrast" countries, Nigeria and Bolivia, is not easily reduced to social networking and literacy levels especially when we also consider the numbers from 2010 (Table 2).

Table 2. Level of social networking and literacy rate, 2010

Country	Facebook subscribers (% of population)	Mobile cellular subscription per 100 inhabitants)	Internet usage (% of users)	Literacy rate(% of population)
Tunisia	27.9	84.9	36.8	88.9
Egypt	13.0	87.1	26.7	66.4
Libya	7.1	171.5	14	97.7
Nigeria	2.8	55.1	28.4	66.6
Bolivia	14.6	72.3	20	90.7

Sources: Internet World Stats: Usage and Population Statistics, 2012; International Telecommunications Union, ITU (2012); United Nations Development Program, UNDP (2012); Nigerian Communication Commission, NCC (2012).

In 2010, Egypt, at 87.1, recorded a higher level of cell phone diffusion than Tunisia, 84.9, (though not by much) and significantly lower than Libya (171.5). It also ranked much lower than Libya in literacy level (66.4 to 97.7). However, Egypt's Facebook population (13.0) and Internet diffusion (26.7) were nearly double those of Libya. These numbers could support the argument that social networking contributed to the relatively non-violent (compared to the massive bloodshed in Libya) resolution of the Egyptian conflicts (though much of the issues that led to the protests remain unresolved). The data might also explain Bolivians' success in reversing the two policies against which they protested: increase in the price of gasoline and the construction of an international highway. As Table 2 shows, Bolivia rates highly on all the indicators. One could also argue that the poor penetration level of Internet access in Nigeria and low number of Nigerians on Facebook explain the ease with which the Jonathan administration subdued the #OccupyNigeria movement, though not before at least 20 people had been fatally shot by Nigeria's military. Also, with less than four percent of the 160 million Nigerians on Facebook, social networking could only play a limited role in the protests.

We explore these possibilities but draw no conclusions without further research. What is clear at this point, as Fisher (2010) notes, is that the people, not technology, were crucial in the mass protests that occurred in the five countries that we have considered. "In Tunisia, anger brought people to the street. They could look in the face of people next to them and saw people who felt what they too felt. After all, organizing through social media also draws the attention of government officials. In Egypt, there were 70 million people but only 20,000 were on Twitter at the time" (Fisher 2010). While the 20,000 people on Twitter and the 13% of the population on Facebook between 2010 and 2011 did internationalize the crisis in Egypt, the critical work was done by the thousands who amassed for three weeks at Tahrir Square in Cairo.

We note a common thread in the three North African countries involved in the Arab Spring: the presence of heads of state who had run the country for a long time, thus constituting dictatorships. While Libyans were most certainly socioeconomically better off than other Africans on the continent, they evidently harbored a simmering discontent about Colonel Muammar Gaddafi's 42-year regime. In Tunisia, though the Ben Ali Administration had been in power for only 12 years, there were also deep-seated grievances that erupted when a street vendor protested the oppressive practices of local government officials. Mubarak had been Egyptian president for 30 years in a country that experienced regular religious, political and social turbulence. Conversely, Nigeria and Bolivia were countries experiencing some level of democracy, as evaluated by the fact that general elections had been held in recent years. In the case of Nigeria, President Jonathan was in his second year of his first full term (having stepped in to complete late President Musa Yar'Adua's term). Thus, political bargaining between the protesters and the government, regardless of the means of communication, occurred in a somewhat more dynamic and functional political environment.

CONCLUSION

This analysis addresses some of the questions with which we started this chapter: what precisely did social networking do in the Arab Spring, and can we draw conclusions about the trajectory of such mass protests in other countries? Does the use of social networking transform political mobilization in any unique way or are they merely tools to achieve political goals in ways that provide continuity with previous means of communication such as the electronic and print media? In many African countries, the electronic media were owned and operated by the government. They therefore served mostly as channels for official government information. Indeed, the electronic media in Nigeria before the 1990s were branches of federal and state ministries of information. There were no distinctions between them and the functions of the supervising information ministries. Radio, though a mass medium of communication, became a tool to serve the interest of the power elites. Even the printing press noted as the technology that set the stage for the modern state system was initially an instrument of the powerful. While it did accelerate the Reformation and the events that led to the modern state system, it was not until the 1800s that there was universal education in Europe (Cordasco, 1976). Before then therefore, the masses still depended on the elite for access to the written material. In the case of radio in Nigeria, even after the deregulation of the sector and private/commercial radio and television competed for airwaves with government-operated electronic media, there was rarely much opposition programming or messages. Rather, the privately owned electronic media became channels of entertainment and advertising. Indeed, as the radio licenses were given on the FM spectrum, "FM radio" soon became synonymous with entertainment.

Still, broadcasting is not exactly like social networking. The latter departs from the traditional conception of electronic communication in its lack of hierarchy and authority. As Urwin (2012) argues, with social networking, power has devolved to the individual – "those wonderful hackers, those stealing the wonderful programs that have been developed by the powerful manufacturers. We can hack into strange systems that oppress us." Social networking gives people power – but it also empowers the already powerful. For instance, while 34.3% of the world population are Internet users, less than half, 15.6% of these are Africans, and 42.9% are in Latin America and the Caribbean (Internet World Stats, 2013). If social networks are empowering, they also increase the disempowerment of those on the global margins and digitally unconnected and disconnected. Yet the indigenous protesters in the Bolivian Amazon were able to mobilize and march without being very connected. Undoubtedly, social movements and protests against governments are not new phenomena, and they have occurred throughout history regardless of the state of technological development.

Social movements and political mobilization utilize available means of communication; different technologies have had different impacts depending upon the era and the context. Information and communication technologies, however, could change the context in the future if more of the disempowered gain access and can utilize the empowering potential of the technologies. Even with this individual empowerment, however, states will most likely still exert power within the networks, and causes of mass mobilizations will still need to be sought within the physical social and political environments. We therefore conclude with the words of Adam Salkheid (2012): "Let's not get carried away. The Arab Spring, Facebook revolution makes for a good sound bite that can be framed in 140 characters. Complex social change can't be summed up in 140 characters or three second sound bites." Neither does a hash tag (as in #Occupy Nigeria) make a revolution.

REFERENCES

Akpan-Obong, P. (2009). Information and Communication Technologies in Nigeria: Prospects and Challenges for Development. New York: Peter D. Lang.

Akpan-Obong, P. (2011). This technology of good and evil. Saturday Punch. October 22, 2011.

Ameh, G. (2013). Four dead as Emir of Kano escapes assassination attempt. Daily Post. January 19, 2013. http://dailypost.com.ng/2013/01/19/four-dead-as-emir-of-kano-escapes-assassination-attempt/

Brinkerhoff, J. (2009). Digital Diasporas: Identity and Transnational Engagement. Cambridge: Cambridge University Press.

Cameron, M. & Hershberg, E. (2010) Latin America's Left Turns: Politics, Policies and Trajectories of Change. Boulder: Lynne Reinner.

Castells, M. (1996). The Information Age-Economy, Society and Culture, Vol 1: The Rise of the Network Society. Oxford: Blackwell Publishers.

CNN News Wire Staff (January 09, 2012). Nigerians protest end of fuel subsidy. http://articles.cnn.com/2012-01-09/africa/world_africa_nigeria-strike_1_nigeria-labour-congress-protests-and-strikes-fuel-subsidy?_s=PM:AFRICA.

Cordasco, F. (1976). A Brief History of Education: A Handbook of Information on Greek, Roman, Medieval, Renaissance, and Modern Educational Practice. Rowman & Littlefield.

Dartnell, M. (2006). Insurgency Online: Web Activism and Global Conflict. Toronto: University of Toronto Press.

Deibert, R. & Rohozinski, R. (2010). Liberation vs. Control: The future of cyberspace. Journal of Democracy, 21, 4: 43-57.

Diamond, L. (2010). Liberation technology. Journal of Democracy, 21, 3: 69-83.

Falola, T. & Heaton, M. M. (2008). A History of Nigeria. Cambridge University Press.

Fisher, A. (2012). Presentation at a session on Mobiles, Social Media and Democracy, at Atlanta 2012: International conference on information and communication technologies and development, March 12-15, in Atlanta, Georgia.

Flood, A. (Jan. 9, 2012). Chinua Achebe leads Nigerian authors' fuel subsidy protest. Guardian. Retrieved: http://www.guardian.co.uk/books/2012/jan/09/chinua-achebe-nigeria-fuel-subsidy-protests.

Fox News (2010). China Lashes Out at U.S. for Internet Criticism. http://www.foxnews.com/politics/2010/01/22/china-lashes-internet-criticism/#ixzz2Jv9Ekpel.

Garrett, R. (2006). Protest in an information society: A review of literature on social movements and new ICTs. Information, Communication & Society, 9, 2: 202-224.

Ghonim, W. (2012). Revolution 2.0: The Power of the People Is Greater than the People in Power. Houghton Mifflin Harcourt.Golkar, S. (2011). Liberation or suppression technologies? The Internet, the Green Movement and the regime in Iran. International Journal of Emerging Technologies, 9, 1: 50-70.

Goggin, G. & McLellan, M. (eds). (2008). Internationalizing Internet Studies: Beyond Anglophone Paradigms. New York: Routledge.

Grant, R. (2004). The Democratization of diplomacy: Negotiating with the Internet. Oxford Internet Institute, Research Report No. 5. http://ssrn.com/abstract=1325241

Gurr, T. (1970). Why Men Rebel. Princeton, N.J.: Princeton University Press.

Howard, P. (2010). The Digital Origins of Dictatorship and Democracy: Information Technology and Political Islam. Oxford University Press.

Held, D., McGrew, A., Goldblatt, D. & Perraton, J. (1991). Global Transformations: Politics, Economics, and Culture. Palo Alto, CA: Stanford University Press.

Herrera, L. (2012). Egypt's Revolution 2.0: The Facebook Factor. Jadaliyya. http://www.jadaliyya.com/pages/index/612/egypts-revolution-2.0_the-facebook-factor

Howard, P. & Hussain, M. (2011). The upheavals in Egypt and Tunisia: The role of digital media. Journal of Democracy, 22, 3: 35-48

Institute for Homeland Security Solutions (2009). How political and social movements form on the Internet and how they change over time.

Internet World Stats: Usage and Population Statistics (2012). http://www. Internetworld stats.com/

International Telecommunication Union, ITU (2012). ICT Statistics. http://www.itu.int/ITU-D/ict/statistics/

Johnson, C. (2010). Dismantling the Empire. Metropolitan Books.

Kenner, D. (2012) Conisur march to demand road through TIPNIS national park. Bolivia Diary. http://boliviadiary.wordpress.com/2012/01/18/conisur-march-to-demand-road-through-tipnis-national-park/

Lim, M. (2012). Click, cabs and coffee houses. Journal of Communication, 62: 231–248.

Maiangwa, B. & Uzodike, U. O. The changing dynamics of Boko Haram terrorism. Aljazeera Center for Studies. http://studies.aljazeera.net/en/reports/2012/07/ 2012731 6859987337. htm.

Mansour, G. (1993). Multilingualism and Nation Building. Multilingual Matters.

Massey, G. (2012). Ways of Social Change: Making Sense of Modern Times. Thousand Oaks, Ca: Sage Publications.

Mayer-Schonberger, V. & Lazer, D. (2007). Governance and Information Technology: From Electronic Government to Information Government. Boston, MA: The MIT Press.

Millennium Development Goals Database, United Nations (2013). Mobile cellular subscriptions per 100 inhabitants. http://data.un.org/Data.aspx?d=MDG&f=seriesRowID%3A756.

Mossberger, K., Tolbert, C. & McNeal, R. (2007). Digital Citizenship: The Internet, Society, and Participation. Boston, MA: The MIT Press.

Mudhai, O.F. (2009). African Media and the Digital Public Sphere. New York: Palgrave.

Murphy, E. (2006). Agency and space: The political impact of information technologies in the Gulf Arab states. Third World Quarterly, 27, 6: 1059-1083.

Musa, A. O. (2012). Socio-economic incentives, new media and the Boko Haram campaign of violence in Northern Nigeria. Journal of African Media Studies, 4, 1: 111-124.

Nesbitt-Ahmed, Z. (2012). Women and #OccupyNigeria. Gender Across Borders: A Global Voice for Gender Studies. http://www.genderacrossborders.com/2012/02/14/women-and-occupynigeria/

Nigerian Communications Commission, NCC (2012). Industry statistics. www.ncc.gov.ng/

Noor, N. (2011). Tunisia: The Revolution that started it all. International Affairs Review. http://www.iar-gwu.org/node/257.

Odutola, K. (2012). Diaspora and Imagined Nationality: USA-Africa Dialogue and Cyberframing Nigerian Nationhood. Durham, NC: Carolina Academic Press.

Ojeleye, O. (2010). The Politics of Post-war Demobilisation and Reintegration in Nigeria. London, UK: Ashgate Publishing, Ltd.

Olofinlua, T. (2012). Nigerians Wield Social Media During Fuel Subsidy Protests. Global Press Institute. http://www.globalpressinstitute.org/global-news/africa/nigeria/nigerians-wield-social-media-during-fuel-subsidy-protests#ixzz1qFcdQ0va.

Omoruyi, O. (1999). The Tale of June 12: The Betrayal of the Democratic Rights of Nigerians (1993). Press Alliance Network Limited.

Ori, K. A. (2009). An informed evaluation of the Nigerian Civil War of 1967: A social science case study." http://www.nigerianmuse.com/20091215100026zg/sections/general-articles/an-informed-evaluation-of-the-nigerian-civil-war-of-1967-a-social-science-case-study/

Rivera, P. (2011) "Politicos Bolivianos en Twitter." Government in the Lab. http://govinthelab.com/julio-politicos-bolivianos-en-twitter/.

Road Rage: The Splintering of Evo Morales's Base. The Economist, October 1, 2011. http://www.economist.com/node/21531009.

Romero, S. (2011) After move to cut subsidies, Bolivians ire chastens leader." New York Times, January 30, 2011. http://www.nytimes.com/2011/01/31/world/americas/31bolivia.html.

Salkheid, A. (2012). Presentation at a session on Mobiles, Social Media and Democracy, at Atlanta 2012: International conference on information and communication technologies and development, March 12-15, in Atlanta, Georgia.

Serberny, A. & G. Khiabany (2010). Blogistan: The Internet and Politics in Iran. London: Tauris.

Snow, D, Cress, D., Downey, L., & Jones, A. (1998). Disrupting the quotidian: Reconceptualizing the relationship between breakdown and the emergence of collective action. Mobilization: An International Journal 3: 1-22.

Snow, D. & Soule, S. (2010). A Primer on Social Movements. W.W. Norton.

Skocpol, T. (1979). States and Social Revolutions. Cambridge, UK: Cambridge University Press.

Tilly, C. (1984). Big Structures, Large Processes, Huge Comparisons. Russell Sage Foundation.

United Nations Development Program, UNDP (2012). Research and publications. http://www.undp.org/content/undp/en/home/librarypage.html.

United States Institute of Peace. (2010). Can you help me now? Mobile phones and peacebuilding in Afghanistan? http://www.usip.org/events/can-you-help-me-now-mobile-phones-and-peacebuilding-in-afghanistan.

Urwin, T. (2012). Presentation at a session on Mobiles, Social Media and Democracy, at Atlanta 2012: International conference on information and communication technologies and development, March 12-15, in Atlanta, Georgia.

Uwechue. R. (2004). Reflections on the Nigerian Civil War: Facing the Future. Trafford Publishing.

VozBoliviana.blogspot.org. (2012) Amigos o enemigos? http://vozboliviana.blogspot.com/

Van Laer, J. (2007). Internet use and protest participation: How do ICTs affect mobilization? PSW-paper 2007/1. http://webhost.ua.ac.be/m2p/publications/PSWPaper2007_1_ Jeroen VanLaer.pdf.

Wehrenfennig, D. (2006). Beyond diplomacy: Conflict management in the network society. Paper presented at the 2006 annual meeting of the American Political Science Association, in Philadelphia. http://www.allacademic.com/meta/p151460_index.html.

In: Social Networking
Editors: X. M. Tu, A. M. White and N. Lu
ISBN: 978-1-62808-529-7

Chapter 3

RELATIONSHIPS BETWEEN PERSONALITY AND INTERACTIONS IN FACEBOOK

Fabio Celli*[1]*and Luca Polonio*[2]†**
[1]CLIC-CIMeC, University of Trento
[2]DiSCoF, University of Trento

Abstract

In this paper we address the issue of how users' personality affects the way people interact and communicate in Facebook. Due to the strict privacy policy and the lack of a public timeline in Facebook, we automatically sampled data from the timeline of one "access user". Exploiting Facebook's graph APIs, we collected a corpus of about 1100 ego-networks of Italian users (about 5200 posts) and the users that commented their posts. We considered the communicative exchanges, rather than friendships, as a network. We annotated users' personality by means of our personality recognition system, that makes use of correlations between written text and the Big5 personality traits, namely: extroversion, emotional stability, agreeableness, conscientiousness, openness. We tested the performance of the system on a small gold standard test set, containing statuses of 23 Facebook users who took the Big5 personality test. Results showed that the system has a average f-measure of .628 (computed over all the five personality traits), which is in line with the state of the art in personality recognition from text. The analysis of the network, that has a average path length of 6.635 and a diameter of 14, showed that open-minded users have the highest number of interactions (highest edge weight values) and tend to be influential (they have the highest degree centrality scores), while users with low agreeableness tend to participate in many conversations.

Keywords: Social Network Analysis, Personality Recognition, Facebook, Natural Language Processing

*E-mail address: fabio.celli@unitn.it
†E-mail address: luca.polonio@unitn.it

1. Introduction

1.1. Overview

Written text convey a lot of information about the personality of its author (Mairesse et al., 2007 [22], Argamon et al., 2005 [2]), and today machine learning techniques allow us to extract personality from text automatically, with a certain degree of accuracy. Online Social Networks (OSN) are huge repositories where user-generated written texts (posts) are found associated with their authors (users), hence they are the perfect place for the automatic extraxtion of personality and the analysis of how it affects interactions among users.

Facebook in particular is a social network service launched in 2004 that allows any person, who declare themselves to be at least 13 years old, to become registered users. In 2012, Facebook counted over 900 million active users, who can create personal profiles and communicate with friends and other users through private or public messages and receive information about their friends by means of a news feed timeline. To allay concerns about privacy, Facebook enables users to choose their own privacy settings and choose who can see specific parts of their profile. Only a user's name and profile picture are required to be accessible by everyone. The rest of the information on users' pages are by default visible only to friends or to friends-of-friends. Boyd and Hargittai 2010 [5] highlight that while news media were critic toward the company's privacy policies, Facebook has continued to attract more users to its service, indicating that people cares a lot about privacy issues. Another interesting study (Bunloet et al., 2010 [7]) over more than 7000 students in Thailand, showed that two top reasons why people use Facebook are 1) having conversation with friends and 2) reducing stress.

In recent years there has been a great effort in the analysis of OSN, and there has been a great interest toward the analysis of Facebook in particular (see for example Catanese et al., 2011 [8]) because, despite it is very challenging to extract data from it for its privacy policy, it is one of the largest and general purpose existing social networks online.

Personality Recognition from Text (PRT henceforth) consists in the automatic classification of authors' personality traits from pieces of text they wrote. This task, that is partially connectecd to authorship attribution, requires skills and techniques from Linguistics, Psychology, Data Mining and Communication Sciences. For example PRT requires some correlations between language features and personality traits (provided by psychologists), a solid background in Data Mining for classification, a good knowledge of communication practices for the social analysis and, most important, a formalized personality schema in order to define classes.

PRT in social networks online is a really challenging task: posts are often very short and noisy, and normal tools for Natural Language Processing (NLP) often perform bad online (Maynard et al., 2012 [23]). In addition the strict privacy policies of many social network services, included Facebook, put heavy restrictions on data sampling. In this work we analyse a network of Italian Facebook users related by their communicative exchanges, for instance posts and comments, rather than by friendship as other social network analysts did. For example Quercia et al., 2012 [28] tested the hypothesis that people having many social contacts on Facebook are the ones who are able to adapt themselves to new forms of communication, present themselves in likable ways, and have propensity to maintain superficial relationships, but they found that there is no statistical evidence to support such a conjecture.

We are going to study how users' personality affects the way people interact and communicate in a OSN, rather than study the effect of personality on friendship connections. To do so we developed and tested a personality recognition system. In a previous work [10], we analysed the effects of one specific personality trait, emotional stability, in Twitter. Here we aim to go further in the research including all the five personality traits provided by the Big5, namely: extroversion (e), emotional stability (s), agreeableness (a), conscientiousness (c) and openness to experience (o).

The paper is structured as follows: in the remainder of this section we will provide an introduction to personality in psychology, to the Big5 and to previous and related work. Then, in the next sections, we will provide a description of the personality recognition system, how we tested its performance on Facebook data and we will introduce how we collected the Facebook dataset. In the end we will report and discuss the results of the experiment.

1.2. Personality

According to psychologists (DeYoung 2010 [14]) and neuroscientists (Adelstein et al., 2011 [1]), personality is defined as an affect processing system that describes persistent human behavioural responses to broad classes of environmental stimuli. It characterises a unique individual and it is involved in communication processes and connected to how people interact one another.

The Standard Way to formalize personality in psychology is the Big5 factor model, introduced in by Norman in 1963 [24]. It emerged from empirical analyses of rating scales, and has become a standard over the years. The five bipolar personality traits, namely Extroversion, Emotional Stability, Agreeableness, Conscientiousness and Openness, have been proposed by Costa & MacCrae 1985 [13].

Extroversion is bound to energy, positive emotions, surgency, assertiveness, sociability and talkativeness. Emotional stability is bound to impulse control, and is sometimes referred by its low pole: neuroticism that is the tendency to experience unpleasant emotions easily, such as anger, anxiety, depression, or vulnerability. Agreeableness refers to the tendency to be compassionate and cooperative rather than suspicious and antagonistic towards others. Conscientiousness is the tendency to show self-discipline, act dutifully, and aim for achievement; planned rather than spontaneous behaviour, organized, and dependable. Openness to experience is bound to the appreciation for unusual ideas, to curiosity, and variety of experience. It often reflects the degree of intellectual curiosity, creativity and a preference for novelty and variety.

According to Digman 1990 [15], there has been a lot of studies in psychology that independently came to the conclusion that five are the right dimensions to describe personality. Despite there is a general agreement on the number of traits, there is no full agreement on their meaning, since some traits are vague. For example there is some disagreement about how to interpret the openness factor, which is sometimes called "intellect" rather than openness to experience.

The Big5 has been replicated in a variety of different languages and cultures, such as

Chinese (Trull & Geary 1997 [31]) and Indian (Lodhi et al., 2002 [19]). Some researchers, such as Bond et al., 1975 [4] and Cheung et al., 2011 [11] suggest that the openness trait is particularly unsupported in asian cultures such as Chinese and Japanese, and that a different fifth factor is sometimes identified. Also the relationship between language and personality has been investigated (Mairesse et al., 2007 [22]), although there are very few applications for personality recognition in languages different from English. This is a good reason for experimenting with personality recognition in Italian.

1.3. Previous and Related Work

There are two main disciplines that are interested in the extraction of personality from OSN: one is computational linguistics, that extracts personality from text, and the other one is the community of social network analysts, that extract information about personality from network configuration (see for example [30]) as well as from other extralinguistic cues (see Bai et al., 2012 [3]).

The computational linguistics community became interested in PRT first. In 2005 a pioneering work by Argamon et al. [2] classified neuroticism and extroversion using linguistic features such as function words, deictics, appraisal expressions and modal verbs. Oberlander & Nowson 2006 [25] classified extroversion, stability, agreeableness and conscientiousness of blog authors' using n-grams as features and Naive Bayes (NB) as learning algorithm. Mairesse et al., 2007 [22] reported a long list of correlations between Big5 personality traits and the features contained in two external resources: LIWC (see Pennebaker et al., 2001 [27] for details) and RMC (see Coltheart 1981 [12] for details). The former includes features such as word classification, like "positive emotions" or "anger", while the latter includes scores like age of acquisition of word or word imageability. They obtained those correlations from psychological factor analysis on a corpus of Essays (see Pennebaker & king 1999 [26] for details) and developed a supervised system for personality recognition[1].

Luyckx & Daelemans 2008 [20] built a corpus for stylometry and personality prediction from text in Dutch using n-grams of Part-Of-Speech and chunks as features. They used the Myers-Briggs [6] Type Indicator schema, that includes 4 binary personality traits, in place of the Big 5. Unfortunately their results are not comparable to any other because of the different language and personality schema used. In a recent work, Iacobelli et al., 2011 [17] tested different features, such as stop words or inverse document frequency. They found that word bigrams with stop words treated as boolean features yield very good results for predicting personality in a large corpus of blogs using Support Vector Machines (SVM) as learning algorithm. As is stated by the authors themselves, their model may overfit the data, since the bigrams extracted are very few in a very large corpus. Kermanidis 2012 [18] followed Mairesse 2007 and developed a supervised system based on low level features, such as Part-of-Speech and words associated to psychological states like in LIWC. She trained a SVM classifier on Modern Greek, obtaining good results and demonstrating that correlations between traits and language can be successfully ported to other languages.

Most of the computational linguists used accuracy (acc) as evaluation measure, but recently there is a shift towards f-measure (f), that is representative not just of the precision of

[1] demo available online at http://people.csail.mit.edu/francois/research/personality/demo.html

a system , but also of its coverage and replicability. Among social network analysts, Golbeck et al., 2011 [16], predicted the personality of 279 users from Facebook using either linguistic (such as word count) and social network configuration features (such as friend count). Using M5 trees as learning algorithm (M5), they predicted personality trait scores rather than classes, and reported mean absolute error (mae) as evaluation measure. A summary of the works in personality recognition is reported in table 1. Although the results of different scholars are not directly comparable one another because they used different datasets and different evaluation metrics, we can see that there has been an increase in performance in recent years, which went hand in hand with the renewed interest in PRT.

Table 1. summary of PRT.

Author	Alg.	Measure	Traits	Results (avg).
Argamon 2005	NB	acc	es	0.576
Oberlander 2006	NB	acc	esac	0.539
Mairesse 2007	SVM	acc	esaco	0.57
Iacobelli 2011	SVM	acc	esaco	0.767
Golbeck 2011	M5	mae	esaco	0.115
Celli 2012	-	pacc	esaco	0.631
Kermanidis 2012	SVM	f	esaco	0.687

All the systems described adopt a supervised approach to PRT. This means that they retrieve a model from a finite, labeled set of data by using machine learning techniques, and then apply those models to other, larger, datasets. We suggest that the supervised approach in PRT has some major limitations due to 1) a high risk of overfitting; 2) poor domain and language portability and 3) great difficulties in the annotation of data for training.

We presented the first system for PRT that does not require supervision in Celli 2012 [9]. We exploited correlations between linguistic features and personality traits adapted from Mairesse et al., 2007 for the prediction of personality in Italian. The advantage of this system is that it automatically adapts to the data at hand, avoiding the risk of overfitting and raising domain and language adaptation. The drawback is that it requires a small labeled set to evaluate the results post-hoc or to estimate accuracy (pacc), as we did in [9]. We report the details about our system in the next section.

2. Automatic Personality Recognition on Facebook Data

2.1. Description of the System

The system for PRT takes as input 1) unlabeled text data with authors; 2) a set of correlations between personality traits and linguistic features: we used the one reported in table 2. We picked up from Mairesse et al., 2007 the correlations with cross-language features, for instance: punctuation, exclamation marks, parentheses, question marks, quotes, word repetition ratio and average word frequency. The system has two outputs: 1) one model of personality for each author and 2) a confidence score for each personality trait of models generated. This is based on the assumption that one user has one and only one complex personality, and that this personality emerges at various levels from written text as well as from other extralinguistic cues.

Table 2. Features and correlations with personality traits.

feat.	extr.	em. st.	agree.	consc.	open.
punct.	-.08**	-.04	-.01	-.04	-10**
excl. marks	-.00	-.05*	.06**	.00	-.03
numbers	-.03	.05*	-.03	-.02	-.06**
parenth.	-.06**	.03	-.04*	-.01	.10**
quest. marks	-.06**	-.05*	-.04	-.06**	.08**
quotes	-.05*	-.02	-.01	-.03	.09**
repeat. ratio	-.05**	.10**	-.04*	-.05*	.09**
avg w. freq.	.05*	-.06**	.03*	.06**	.05**

Personality models are formalized as 5-characters strings, each one representing one trait of the Big5. Each character in the string can take 3 possible values: positive pole (y), negative pole (n) and omitted/balanced (o). For example a "ynoon" stands for an extrovert neurotic and not open minded person. Figure 1 represents the pipeline of the system. In the

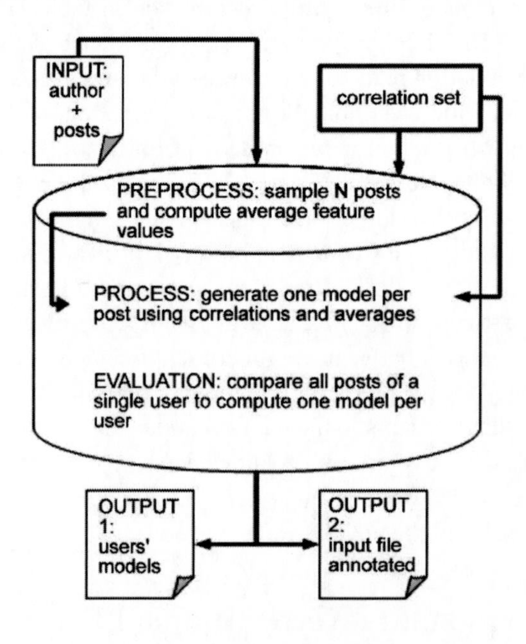

Figure 1. Personality Recognition System pipeline.

preprocessing phase the system samples a small portion of unlabeled data, (the amount can be defined by the user) and analyses the distribution of the features of the correlation set in portion of unlabeled data. With the information about the average feature usage in the dataset we are able to apply filters: for example if a user is found to use more punctuation than the average, the correlation with the punctuation fires, and increase or decrease the score associated to one or more personality traits. This strategy is useful in order to fit the correlation firing to the data, thus avoiding the portability problem.

In the processing phase the system generates one model for each written text, checking for matches of linguistic features provided in the correlation set. If it finds a feature value above the average the system increments or decrements the score associated to the personality trait, depending on a positive or negative correlation. From positive and negative trait scores the system can compute trait confidence scores (tc), defined as $tc = (y - n)$, where $y = \frac{ym}{P}$ and $n = \frac{nm}{P}$. Here ym is the count of matches for the positive pole of personality trait and nm is the count of matches for the negative pole of the trait. P is the count of posts.

In the evaluation phase the system compares all the models generated for each single post of each user and retrieves one model per user. This is based on the idea that, even if a user can express different aspects of personality in different posts, motivated by different goals and situations, we can still catch the personality that the user expresses most of the times by comparing all the posts. Then the system turns the personality scores into classes (if below 0 predicts a negative pole, if above 0 the positive one if it is equal to 0 predicts a "o"). In the evaluation phase the system also computes average confidence and variability for each user. Those measures are computed from the comparison of all models generated from each user's texts. Average confidence (c) gives a measure of the robustness of the personality model. It is defined as $c = \frac{tp}{M}$ where tp is the count of personality models matching within the same user (for example "y" and"y", "n" and "n", "o" and "o") and M is the total of the models generated for that user. Variability gives information about how much one user tends to write expressing the same personality traits in all the posts. It is defined as $v = \frac{c}{P}$ where c is the confidence score and P is the count of all user's texts. Most important: the system can evaluate personality only for users that have more than one post, the other users are discarded.

2.2. Testing the Personality Recognition System

We collected a small test set of Facebook data annotated with personality of users. To do so we run an experiment with 35 participants, who took the Big5 personality test. We asked the participants to leave the URL of their Facebook personal page and to write a short essay, minimum 15 lines and maximum 30, on any subject. The participants are all Italian native speaker students aged between 19 and 27, 10 males and 25 females. A couple of them are bilingual Italian-German speakers.

With their consent, we manually collected their public statuses from participants' Facebook pages. We sampled only text, discarding any other type of data. We could sample data only for 23 of them, which we actually put in the dataset. We built 2 datasets, one with Facebook data and the other one with the data from the essays. We produced the gold standard personality models for the users from the results of the Big5 test. We converted the scores of the big5 into the 3-class format used by the system. To do so we turned the scores above 50 into "y" and all the scores below or equal to 50 into "n".

We run the system on both the datasets, using the same features. Results, reported in table 3, reveal that the system achieves the best performance on Facebook data, even if it

Table 3. PRT on essays (off) and Facebook (fb) text data.

feature set	avg P	avg R	avg F
off	.45	.7	.558
fb	.547	.735	.628

exploits correlations extracted from essays in English. This means that the way the system uses correlations is adaptable and suitable for social network data.

3. Collection of the Dataset

Sampling data from Facebook is hard. This is due to different factors, like the lack of a public timeline and the strict privacy policy. The former factor prevents from sampling data from users of which we do not have the friendship. The latter factor prevents from having access to the information of users with which we do not have the friendship.

The sampling pipeline can be seen in figure 2. We developed a crowler that exploits

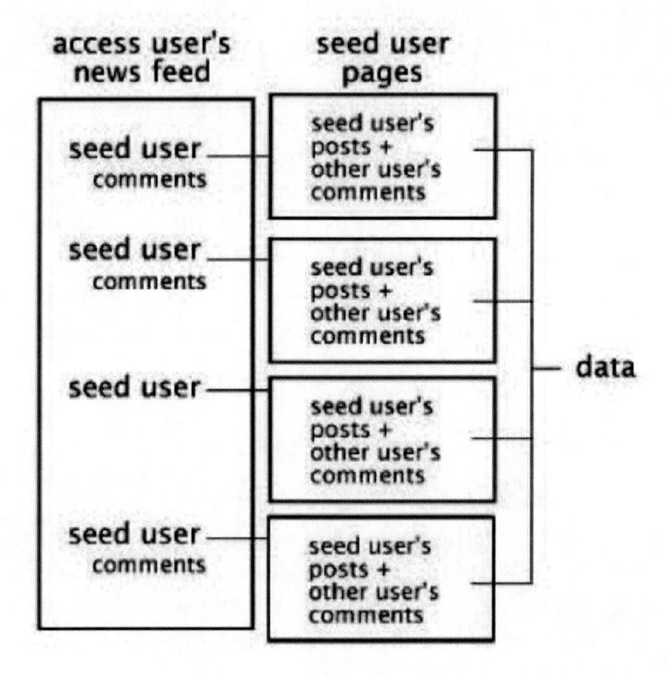

Figure 2. Sampling pipeline.

Facebook's graph API[2] in order to sample users' statuses. The system starts from the news feed of a "access user", who suscribed onto Facebook developer and can take the "access token" key for the API. From the timeline of the access user the system extracts some "seed users" and samples all the statuses and comments written either by the seed users and by the "related users" who interacted with them. The system collects a minimum of 2 posts or comments per user and keeps track of all the users' IDs sampled, in order to avoid duplicates. Finally we filtered out groups and fanpages and we kept only users.

The resulting dataset contains the egonetworks of the seed and related users, as depicted in figure 3. Seed users are linked to the related users with weighted "communicative exchanges" relationships. This means that the more a related user commented a seed user,

[2] http://developers.facebook.com/tools/explorer

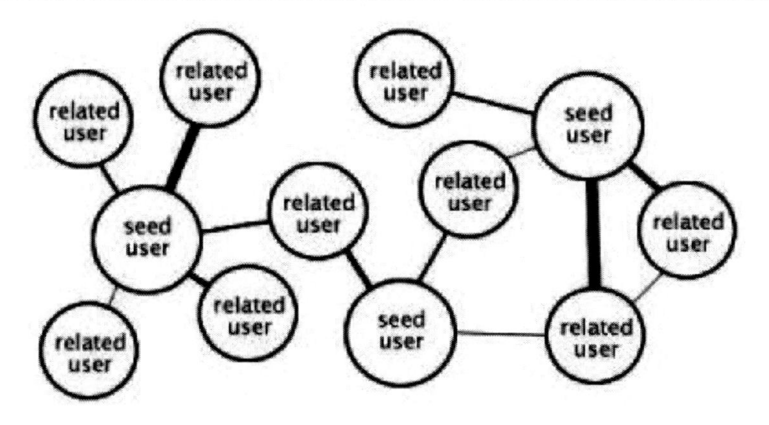

Figure 3. Egonetworks in the dataset.

the more the communicative relationship is considered strong. Although most works use friendships as network connections, we decided to use communicative exchanges because we are interested in the relationship between personality and communicative interactions. In the dataset there are more than 5000 posts and 1100 users. We annotated the personality of each user by means of our personality recognition system.

4. Experiments and Discussion

First of all we retrieved some statistics about the distribution of personality traits in the network and about its topology. The network has a diameter of 14, an average path length of 6.635, average degree centrality of 2.175 and average clustering coefficient of 0.017. This indicates that it is a small network where users have on average a couple of comment-relations each one and with low clustering level. Centrality measures and clustering coefficient have skewed distributions, meaning that a few users have high values and most of them have very low values. The distribution of personality traits, reported in table 4, highlights the low number of extroverted, mentally closed and uncooperative people in the

Table 4. Distribution of personality traits in the network.

trait	y	o	n
extr.	6.2%	66.4%	27.4%
em. st.	13.7%	49.9%	36.4%
agree	31.9%	65%	3.1%
consc.	13.2%	50.4%	36.4%
open.	27.9%	62.2%	9.9%

network. We suggest that this might be due to the personality of the access user ("noyyy"), that influences the selection of people who are in the network. We will refer to this problem as the "access user bias", that is related to the sampling procedure and does not take place in those networks, like Twitter, where there is a public timeline available.

We analysed the relationship between personality and interactions by computing the association between personality traits and some topology measures, like degree centrality, correlation coefficient and edge weight. In order to do that we measured association scores by computing $as = \frac{bti}{td}$, where *bti* are the 10 most frequent personality traits associated to each topology measure used, and *td* is the trait distribution reported in table 4.

Results, reported in table 5, show several interesting phenomena. First of all that in-

Table 5. Association scores.

degree centr.	extr.	em. st.	agree.	consc.	open.
y	0.774	1.387	2.687	2.167	**3.244**
o	0.215	0.381	0.22	0.472	0.077
n	**2.956**	1.701	0	1.308	0.485
edge weight	extr.	em. st.	agree.	consc.	open.
y	0	2.335	2.351	1.923	**3.405**
o	0.15	0.2	0.307	0.198	0.08
n	**3.284**	0.364	1.61	1.785	0
clustering c.	extr.	em. st.	agree.	consc.	open.
y	1.396	0.912	1.567	0.477	1.57
o	0.848	0.501	0.674	0.869	0.905
n	1.016	1.717	**2.032**	1.374	0

troverted and open minded users have the highest degree centrality in the network. In other words they are the ones that are more central and more prone to catch conversations. It is not a surprise that open minded users are in this position, but it is very interesting to note that introvert people have a high degree centrality score too. A closer look to the data reveals that the open minded and introvert traits come often together in the dataset. We suggest this might be due again to the access user bias, as we found previous work [10] that there is a general tendency to have conversations between users that share the same traits. The highest edge weight scores are again associated to open minded and introverted users. This means that those users have the strongest links, in other words the highest number of comments. We interpret this as a consequence of the position those users occupy in the topology of the network. Also Agreeable and emotionally stable users have high degree centrality and edge weight scores, indicating that those personality traits play a role in being influential in a conversation network. The distribution of high edge weights is very skewed: there are very few strong links and really a lot of links with low weight.

The personality trait associated to high clustering coefficient scores is low agreeableness. If clustering coefficient is related to users' connectedness and links represent comment relationships here, we can interpret this fact as a hint that uncooperative users tend to participate in many conversations in order to debate in a polemic way. The distribution of clustering coefficient scores is very skewed too.

The outcomes of this experiment show the behaviour of a network of interacting users visible to the access user, hence are not generalizable, but yet interesting to study the role that personality traits play in social interactions in a micro network.

5. Conclusion

In this work we have sampled a network of communications between Italian users in Facebook, sampled from one access user's timeline. We automatically annotated it with personality traits in order to analyse how people's personality affects interactions online. We tested the system used for the annotation on a gold standard, obtained from 23 Italian Facebook users which took the Big5 personality test, and provided us with the public data from their timelines. The system proved to label correctly 62% of the data.

From the analysis of the most frequent traits associated to topology measures like degree centrality and correlation coefficients, emerged that open minded and introvert users have the highest degree centrality and the strongest links. We interpreted this evidence as introvert and open minded users (those traits come frequently together in the dataset) tend to be very interested to the information that passes through the network, and tend to post interesting (high commented) statuses. Another interesting result is that the users that have high correlation coefficient have low agreeableness. We interpreted this fact as as a hint that uncooperative users tend to participate in many conversations in order to debate in a polemic way.

The results show how people's personality can be successfully analysed with a quantitative approach on large scale data, yielding very interesting findings. It is not easy to interpret the results, but the same difficulty is found in much of the quantitative sociology based on big data [29]. We suggest that pairing personality recognition with sentiment analysis or topic extraction would make it more informative and easier to interpret. We also suggest that the comparison of personality recognition in communication exchanges and friendship relations, for example using the multi-layer model proposed by Magnani & Rossi 2011 [21], would bring out useful information.

The access user bias, that is due to the restrictions imposed by Facebook and to the lack of a public timeline, prevents from the generalization of those results. Yet it is interesting to observe that a micro network is filtered by the access user according to personality, among other factors. This underlines one more time the importance of personality recognition in the study of social networking.

References

[1] Adelstein J.S, Shehzad Z, Mennes M, DeYoung C.G, Zuo X-N, Kelly C, Margulies D.S, Bloomfield A, Gray J.R, Castellanos X.F, Milham M.P. (2011). Personality Is Reflected in the Brain's Intrinsic Functional Architecture. In *PLoS ONE* 6(11). pp. 1–12.

[2] Argamon, S., Dhawle S., Koppel, M., Pennebaker J. W. (2005). Lexical Predictors of Personality Type. In *Proceedings of Joint Annual Meeting of the Interface and the Classification Society of North America*. pp. 1–16.

[3] Bai.,S. Zhu.,T. Cheng.L. 2012. Big-Five Personality Prediction Based on User Behaviors at Social Network Sites. In *eprint arXiv:1204.4809*. Available at http://arxiv.org/abs/1204.4809v1.

[4] Bond, M.H. Nakazato, H.S. Shiraishi, D. (1975). Universality and distinctiveness in dimensions of Japanese Person Perception. In *Journal of Cross-Cultural Psychology*. 6. pp.346–355.

[5] Boyd, D. Hargittai, E. (2010). Facebook Privacy Settings: Who Cares? In: *First Monday* 15 (8). Available online at http://firstmonday.org/htbin/cgiwrap/bin/ojs/index.php/fm/article/view/3086/2589.

[6] Briggs, I. Myers, P.B. (1980). *Gifts differing: Understanding personality type*. Mountain View, CA: Davies-Black Publishing.

[7] Bunloet, A., Saikeaw, K. R., Tengrungroj, M., Nalinthutsanai, N., Mungpooklang, T., Dabpookhiew, P., Winkam, T., Arayasilapatorn, N. Premgamone, A., Rattanasiri, A., Chaosakul, A. (2010). Analysis of Facebook Usage by College Students in Thailand. In *25th International Technical Conference on Circuit Systems, Computers and Communications (ITC-CSCC 2010)*. pp. 107–111.

[8] Catanese, S.A, De Meo P., Ferrara, E., Fiumara, G., Provetti, A. (2011). Crawling Facebook for Social Network Analysis Purposes. In *Proceedings of WIMS '11: International Conference on Web Intelligence, Mining and Semantics ACM*. pp. 1–8.

[9] Celli, F., (2012). Unsupervised Personality Recognition for Social Network Sites. In *Proceedings of ICDS*, pp. 59–62.

[10] Celli, F., Rossi, L. (2012). The role of Emotional Stability in Twitter Conversations. In *Proceedings of Workshop on Semantic Analysis in Social Media, in conjunction with EACL*, pp. 1–8.

[11] Cheung, F. M., van de Vijver, F. J. R., Leong, F. T. L. (2011). Toward a new approach to the study of personality in culture. In *American Psychologist*, Advance online publication. pp. 1–11.

[12] Coltheart, M. (1981). The MRC psycholinguistic database. In *Quarterly Journal of Experimental Psychology*, 33A, pp. 497-505.

[13] Costa, P.T., Jr. McCrae, R.R. (1985). The NEO Personality Inventory manual. In *Psychological Assessment Resources*. pp. 5–13.

[14] DeYoung, C.G. (2010). Toward a Theory of the Big Five. In *Psychological Inquiry*. 21: pp. 26–33.

[15] Digman, J.M. (1990). Personality structure: Emergence of the five-factor model. In *Annual Review of Psychology* 41: pp. 417–440.

[16] Golbeck, J. and Robles, C., and Turner, K. (2011). Predicting Personality with Social Media. In *Proceedings of the 2011 annual conference extended abstracts on Human factors in computing systems*, pp. 253–262.

[17] Iacobelli, F., Gill, A.J., Nowson, S. Oberlander, J. (2011). Large scale personality classification of bloggers. In *Lecture Notes in Computer Science (6975)*, pp. 568–577.

[18] Kermanidis, K.L. (2012). Mining Authors' Personality Traits from Modern Greek Spontaneous Text. In *4th International Workshop on Corpora for Research on Emotion Sentiment & Social Signals, in conjunction with LREC12*. pp. 90–94.

[19] Lodhi, P. H., Deo, S., Belhekar, V. M. (2002). The Five-Factor model of personality in Indian context: measurement and correlates. In R. R. McCrae J. Allik (Eds.), *The Five-Factor model of personality across cultures*. N.Y.: Kluwer Academic Publisher. pp. 227–248.

[20] Luyckx K. Daelemans, W. (2008). Personae: a corpus for author and personality prediction from text. In: *Proceedings of LREC-2008, the Sixth International Language Resources and Evaluation Conference*. pp. 2981–2987.

[21] M. Magnani and L. Rossi. The ML-model for multi-layer social networks. In *Proceeding of 2011 International Conference on Advances in Social Networks Analysis and Mining, IEEE Computer Society*. pp. 5–12.

[22] Mairesse, F. and Walker, M. A. and Mehl, M. R., and Moore, R, K. (2007). Using Linguistic Cues for the Automatic Recognition of Personality in Conversation and Text. In *Journal of Artificial intelligence Research*, 30. pp. 457–500.

[23] D. Maynard and K. Bontcheva and D. Rout. (2012). Challenges in developing opinion mining tools for social media. In *Proceedings of @NLP can u tag usergeneratedcontent?! Workshop at LREC 2012*. pp. 15–22.

[24] Norman, W., T. (1963). Toward an adequate taxonomy of personality attributes: Replicated factor structure in peer nomination personality rating. In *Journal of Abnormal and Social Psychology*, 66. pp. 574–583.

[25] Oberlander, J., and Nowson, S. (2006). Whose thumb is it anyway? classifying author personality from weblog text. In *Proceedings of the 44th Annual Meeting of the Association for Computational Linguistics ACL*. pp. 627–634.

[26] Pennebaker, J. W., King, L. A. (1999). Linguistic styles: Language use as an individual difference. In *Journal of Personality and Social Psychology*, 77. pp. 1296-1312.

[27] Pennebaker, J. W., Francis, M. E., Booth, R. J. (2001). *Inquiry and Word Count: LIWC 2001*. Lawrence Erlbaum, Mahwah, NJ.

[28] Quercia, D., Lambiottez, R., Stillwell, D., Kosinskiy, M., Crowcroft, J. (2012). The Personality of Popular Facebook Users. In *Proceedings of ACM CSCW 2012*. pp 1–10.

[29] Scott, J. (2011). Social Network Analysis: developements, advances, and prospects. In *Social Network Analysis and Mining*, 1(1). pp. 21–26.

[30] Staiano J, Lepri B, Aharony N, Pianesi F, Sebe N, Pentland A.S. (2012). Friends dont Lie - Inferring Personality Traits from Social Network Structure. In *Proceedings of International Conference on Ubiquitous Computing*. pp. 324–334.

[31] Trull, T. J. Geary, D. C. (1997). Comparison of the big-five factor structure across samples of Chinese and American adults. *Journal of Personality Assessment* 69 (2). pp. 324–341.

In: Social Networking
Editors: X. M. Tu, A. M. White and N. Lu
ISBN: 978-1-62808-529-7

Chapter 4

Resilience to Climate and Demographic Change: The Importance of Social Networks

***Kaberi Gayen*[1] *and Robert Raeside*[2]**
[1]Mass Communication and Journalism, University of Dhaka, Bangladesh
[2]Statistics, Edinburgh Napier University, Scotland

Abstract

As economic, political and environmental systems develop, societies face continued pressure to change. At the beginning of the 21st Century two forces have emerged as major challenges to the wellbeing of humanity; these are population ageing and climate change. The interaction of these is of major concern to vulnerable communities. Social networks provide a means of assessing both physical and social capital to enable people to cope and adapt to change. This can occur by the transmission and diffusion of information and knowledge, allowing aid to be secured and norms to be changed. However, by retrenching traditional ideas and beliefs, social networks can act to damage societies; such as by endorsing low status of women and viewing new ideas and methods with suspicion. In this chapter we show, that if social networks can be designed and made sustainable, then a powerful mechanism can be made available to vulnerable communities which will enhance their resilience to major agents of change. We illustrate how these networks might work, by drawing on data collected from an empirical study of public health interventions in rural Bangladesh.

The concept of society and how networks within it can have harmful or beneficial effects has long been known; as has been witnessed by alliances in the development of cities and nations, the development of scientific and engineering triumphs (see Burke Connections), but it has only comparatively recently been formalised into an interdisciplinary area of academic study known as social Network Analysis (SNA). In SNA, theory is constructed, formulated and tested and a language created to allow codification. The first documented application of social network analysis might well have been by Simmel (1908) and the first sociogram was produced and documented by Moreno in the 1930's. SNA then re-entered the "dark ages", according to Freeman (2004), and along with Scott (2000) and Rogers (2003) they led the re–

emergence of SNA. Since the turn of the century the subject has undergone spectacular growth, as measured by methodological evolution, new application fields, journal articles, books and software. While this evolution of the subject occurred, society has moved on and often faces new challenges. The current challenges are evident in the efforts to provide for an ever increasing population, many of whom live marginal subsistence lives. Indeed it is estimated by the UNDP (2013) that 1.75 billion people live in poverty; most of these dwell in rural areas. Their fragile existence is accentuated by two global forces; those of population ageing and climate change.

These forces are acting simultaneously on subsistence dwellers that depend on agrarian systems of production; populations of the least developed countries are particularly vulnerable to climate change (see Mendelsohn et al., 2006). Farmers and fishermen have to adapt to changes in the environment but how do they know what to do? To complicate this must adapt to new work patterns or raise resources to build dams, invest in new varieties of crops or animals, buy fertilisers etc. Populations worldwide are ageing. Once this was thought to be an issue to be faced by developed affluent nations whose fertility had fallen to below replacement but dramatic fertility transitions have occurred in many of the poorer nations of the world. Added to this, life expectancy has generally increased. This means that in rural agrarian communities there are fewer younger people to do the work and also to care for, increasingly numerous, older people. Migration is a further impact on the old in that children, especially young males, are migrating to major urban centres or even overseas. Often migrants send remittances home but, as economies undergo strictures, earning a surplus to send can become difficult and they are not around to provide physical and emotional help and care.

In such subsistence societies, to cope with disruptive change has been very much the role of the traditional family to provide support; little reliance can be made on the state, especially as their economies are frequently not very competitive. In addition, in many areas non-government organisations just do not have the resources to meet their needs. Ageing and migration means that the traditional family and kinship ties are no longer adequate providers of support. Instead many will have to both receive and give help and care to non-family members; this situation has been found to correlate negatively with happiness and wellbeing (Raeside et al, 2009).

For family to be replaced by society in providing care, a cohesive and functional network has to be developed. By identifying and linking with central actors it can provide access to support and care and also allow ideas to diffuse effectively thorough out the community. There are further advantages of tight social networks in that, if some actors have connections to urban centres and even internationally, then ideas can diffuse into the community. These might allow an ideological change that will enable communities to become fitter to cope with the paradigm shifting effects of ageing and climate change.

In this chapter we focus on the situation in rural Bangladesh to, firstly, ascertain the magnitude of challenges facing subsistence agrarian communities and then to explore the role that the analysis of social networks might have in enhancing the wellbeing of these communities. We end the chapter by listing a series of recommendations for local communities, the Bangladesh Government and the international community.

BACKGROUND

Bangladesh, although exhibiting economic growth and development over the last thirty years, remains a very poor and densely populated country (almost 1200 people per square kilometre). In the UNDP human Development Index Bangladesh is 146 out of 187 and the GDP per capita was only $1,900 in 2011; and 49% of the population has a purchasing power parity of less than $1.25 per day with 26.4% of the population estimated to be in severe poverty. Thus the nation has little capacity to ease adaption to change.

Bangladesh, located in North Eastern corner of the Bay of Bengal, is characterised by the rivers of the Ganges (known locally as the Padma), Meghna and the Brahmaputra, which form a mega-delta. As almost two-thirds of the country is less than five metres above sea level, the country is thus very prone to flooding; a situation not helped by human activities such as removing coastal mangrove forests to facilitate prawn farming (Shahidul et al, 2010). The country has for many years been recognised as vulnerable to climate change (Ali 1999, Agrawala 2003) and, according to the Global Climate Risk Index (Harmeling and Eckstein 2012). Bangladesh is the fourth most vulnerable nation to climate change. The fourth report by the Intergovernmental Panel on Climate Change (IPCC) stated that Bangladeshis expected to experience heavier rainfall in the monsoons and that, as a consequence of melting of Himalayan glaciers, higher river flows will occur; when combined with rises in sea levels, flooding is likely. Rainfall is expected to become heavier and more erratic and droughts to increase in frequency.

20 million people in Bangladesh are expected to be displaced by rising sea levels and, in addition to increased flood risk, tropical storms and cyclones also expected to increase in intensity. Climate change particularly threatens the agricultural sector, which is characterised by rural subsistence farming although it only contributes to 20% of the nation's GDP. Almost half of the Bangladeshi workforce is employed in this sector. Levkoff et al., 1995 and Das et al., 2007 have noted (in other locations), that older people in rural Bangladesh face great challenges in avoiding falling into more extreme poverty, hunger, homelessness and abuse, as a result of the geographical contexts in which they are living.

However, Bangladesh, although resource constrained, has risen to the challenge of climate change; this is evidenced, in that Bangladesh is not as vulnerable nation to climate change as, according to the Climate Risk Index, it was in 2010. Bangladesh has embarked upon a number of states and NGO sponsored projects and, it is reported by the World Bank (2013), that Bangladesh is now a world leader in having invested more than $10 billion over the last three decades in flood prevention, coastal embankments, greenbelt projects, cyclone shelters, warning systems and agricultural research to produce crop resilience to increased salinity levels; but the development of resilience has mainly arisen from community based action, as reported in an article by Friedman (2009) in Scientific American in which Rabab Fatima, South Asia representative for the International Organization for Migration, stated: "This country is quite a miracle, I must say." "It's completely people-driven. Despite all natural odds, despite bad politics and bad governance, people don't starve here". Mahmud and Prowse (2012), examining the pre and post interventions around cyclone Aila, showed that corruption severely compromises the success of public works and NGO interventions. In their survey of 278 Bangladeshi households, 64% reported irregularities with NGO interventions. This points to a major problem with state and institutional intervention; not only are the resources scarce, but corruption in Bangladesh is severely limiting the effectiveness of any intervention from above the community level. According to the Transparency International's 2012 Corruption Perceptions Index, Bangladesh is perceived as highly corrupt with a score of 26 out of 100 (100 equates to no corruption) countries, thus, at best, state interventions have to be viewed with scepticism.

Omar Rahman, in the same article, draws comparisons to how Bangladesh civil society, with prompting from the Government, reacted to the population crises of the late Twentieth Century "We have established a record of doing very complex things," he said. "In a traditional, conservative country, to make it acceptable to talk about birth control shows that we are capable of sustaining social change, if we have enough support." Sovacool et al (2012), from research in South Asia including Bangladesh, argued that infrastructure resilient to climate change can only be effective if hardware (technology) can be blended with social cohesion (community cohesion); the local people need to have knowledge and be willing to adapt to the changing environment; it follows that communication programmes are required. Heltberg et al (2008) developed a case for community-based adaptation and social protection; they advanced an agenda for research and showed that knowledge and finance had to be made available to local communities. Adger (2003), from research based in Vietnam, showed that collective action was effective in adapting to climate change, provided that there could be articulation with social capital.

Thus, to attain resilience and sustainability, it appears that people themselves have to become active. This requires support, encouragement and articulation to knowledge; this gives importance to the role that social networks can play. This will be discussed in the next section but, before going on to discuss this role, the other force of change affecting rural dwellers, that of population ageing, will now be outlined.

Since the 1970's the total fertility rate of Bangladesh has fallen from over 7 children per mother to around 2.2, while at the same time, life expectancy at birth has risen by around 55 years in 1980 to 68.9 years in 2011. One consequence of this is that the proportion of the Bangladesh population who are over 60 years of age has risen from 4.7% in 1981 to over 7% in 2011. While not high compared to the proportion in Western societies (over 20%), for a third world country, it is a major rise and is projected to increase to 11.6% by the end of the decade (data taken from the International Database provided by the US Census 2013). This means that there will be a growing number of people requiring at least some care and support, many of whom are in rural areas making a living by subsistence farming; where the opportunity to raise a surplus is non-existent. State welfare is very limited and old age support has traditionally come from family members (see Cain, 1986; Datta and Nugent, 1984). There are fewer of these supports in a situation that is exacerbated by high migration to urban centres and overseas; especially young males. Remittances used to be sent home but, in harsh economic times, it is difficult for these migrants to fund this. A main form of support is co-residence with younger family members (Knodel and Debavalya 1997) but Knodel and Ofstedal (2002) shows that this is declining (The situation of ageing in Bangladesh is discussed further by Rahman and Ali (2009), Farid et al (2011) , Jesmin and Ingram (2011) and Choudhary (2013).

In the Global South the production of 'difference' relating to age is just as real as in western societies; but with the physical, economic and social consequences being constructed in a different cultural context and with much more profound consequences in terms of the spatialities of vulnerability (Findlay, 2005). For example, rhetoric about the need for adaptation to climate change (DFID, 2009) revolves largely around modifying livelihood options for the 'working-age' population; little mention is given to what increased exposure to flooding (Haines et al, 2006, Nicholls et al., 2007) means for older people, who might traditionally have been less actively engaged in livelihood production. Fairness in relation to adaptation, both to environmental and economic change, therefore demands that attention be devoted also to older people (Adger et al, 2006; Falkingham and Namazie 2002).

In Bangladesh the gender dimension is especially acute in considering the impact of ageing. This arises due to the way that Islamic culture and the patriarchal structure of society has favoured major disparity in the respective ages of marriage partners; making it more probable that older women will be widowed for many years, while at the same time inhibiting the likelihood of them drawing support from outside the household. "Widowed and childless women are especially vulnerable in societies where they lack rights of ownership and property is inherited along male lines." (Barrientos 2003). Typically older people, especially women in rural Bangladesh, face problems in drawing on help outside their immediate family home; this arises from cost issues, the difficulty of travelling unaccompanied and distrust of health professionals (Munsur et al., 2010, Afsana 2004, Ahmed et al 2000 and Paul and Rumsey 2002). Community support is often given to the poor, especially to widows in terms of food, clothing and money. This frequently occurs during religious festivals (Help Aged International Asia Pacific Regional Development Centre, 2000). It is well documented that women, especially older women in rural Bangladesh, avoid formal institutional services such as health centres and health professionals, only using these in dire emergencies; often after first consulting traditional healers. Gayen and Raeside (2007, 2010a and 2010b) have shown that women's connectedness in social networks is an important determinant of reproductive health outcomes and that being well connected is associated with feelings of higher well-being and overcoming the trap of "traditionality".

Compounding these effects, are the economic and social structures in which older people are embedded (Lloyd-Sherlock, 2000). Isolated individuals, notably those with limited human capital (for example, those who are illiterate or have only minimal education) find themselves amongst the most disadvantaged in accessing both state and third sector support systems. This is not to say, that efforts have not been made by external agents to overcome these barriers, but even moderately effective poverty alleviation interventions (Matin and Hulme, 2003) have faced major constraints in delivering support to the most isolated members of the society (Andrews and Entwistle, 2010).

It seems that older people in rural Bangladesh may have to rely on non-family members for care and support in old age. Nilsson et al., (2005) are amongst the growing body of researchers who have shown that in Bangladesh older people's quality of life depends on many factors including being healthy, having a good social support network and security of financial situation. However, unlike in Western countries where getting support from non-family members is not perceived as being detrimental to well-being (Leeson 2006; and Raeside et al., 2009), the change to extra familial support in traditional societies is likely to be unpleasant. The role for social networks is clearly to construct a network of community care and support and provide a bridge to professional health care. In the next section we show the protective effect that can develop from social networks but also show that social networks can inhibit progress.

SOCIAL NETWORKS TO ALLEVIATE THE EFFECTS OF CLIMATE CHANGE

It is in the context outlined above that the engagement of older people in social networks has been shown to be important (Valante, 2010 and Granovetter, 1973) and as a bridge to developing critical social capital for older people (Gayen et al (2010). We postulate that via social structures that accessing and mobalising social capital might be compensated from declining human capital and access to resources, (although we do recognise that this is contentious; Bebbington 2004, Bebbington et al., 2004, Fine 2001and 2007, Harris 2001). In the past, older people were firmly embedded in the wider social networks of their families and of their villages, but with the progressive undermining of the livelihood base of many rural villages in Bangladesh, many people of working ages have left to find work in the cities (or further afield). This has, on one hand, partially disrupted access to social networks for older people, while on the other hand producing the distanciation of social relations (Giddens, 1984) in a context where older people were ill-equipped to stay in touch with distant household members (Wenger, 1994, Zanoon et al, 2006). Wang et al., (2006), Licoppe and Smoreda (2005) and Chetly (2006) point out that integrated communications technology (ICT) can provide a bridge to allow contact and mobile communications and internet access helps migrants to remain in contact with their families, transfer knowledge and in the transfer of remittances. However, poverty and educational disadvantage constrain their use, especially in rural Bangladesh where the lack of a stable technology influence is a further inhibitor. A social network analysis (SNA) approach is therefore valuable in research into allowing disadvantaged communities to establish contacts and bridges to wider society to be sustainable communities by facilitating the diffusion of ideas and aid. Social network analysis provides a framework to investigate the effectiveness of organisational and institutional partnerships in reaching people with very different personal and social connections, (Balkundi and Kilduff 2005).

A constraint and influence on the function of the networks is the local geography with settlements often reflecting both family structures and topology such waterways as in the case of Bangladesh. Thus a spatial dimension needs to be incorporated into SNA which has been lacking in many SNA studies until recently (Daragonova, et al., 2012, Doreian and Conti, 2012, Rindfuss et al., 2004, Verdery et al., 2012 and Viry, 2012). These studies point to the increased knowledge that is gained from combining the different forms of data and Verdery et al (2012) illustrates this in rural villages in Thailand.

To understand the coping strategies of older people as they adapt to new vulnerabilities associated with climate change, the loss of social support from younger family members absent through migration and rising instances of old age poverty a theoretical framework has been constructed. This framework has three interrelated concepts, namely ideational change (Cleland and Wilson, 1987; Kincaid, 2000), social interaction (Bongaarts and Watkins, 1996; Montgomery et al 2001) and social networks (Kincaid, 2000; Kincaid et al., 1993; Rogers and Kincaid, 1981; Valente, et al 1997), which form the core of the approach taken to apply social networks to support excluded communities. These theoretical concepts explain the introduction of new behavioural attitudes into a specific environment as transmitted by the interaction of members of a social network (Rogers and Kincaid, 1981) and it is hypothesised to influence the resilience (or lack of resilience) of older people in different contexts. To measure ideation, Kincaid (2000) used a combination of knowledge, beliefs about attributes of the innovation, the degree of discussion with one's family, peers and change agents, and their social approval about the innovation.

Examining "social systems including social networks" is important to understanding decisions as to wither a member of the household should migrate to secure a better livelihood. This is important as historically the economic and environmental drivers prompting livelihood change have usually been slow and impact on wellbeing over a protracted period (Warner et al., 2009), but climate/environmental change is a catalyst which accelerates the pace of change. Account must also be made of the local geography and environment which sets physical influences on how the framework will operate. Kniveton et al (2008) argue that more research is needed to understand the local and regional contexts in which mobility decision making takes place, but with specific attention to how the benefits of mobility, e.g. remittances, (Gupta et al, 2009; Adams, 2006) are weighed against the costs in terms of the increasing isolation of older people left behind by out-migrants. Migration as a coping strategy, not only impacts the older left-behind population at a household level, but also at a wider community level through depopulation undermining the wider social base on which villages organise themselves (Johnson and Krishnamurthy, 2010; Lloyd-Sherlock, 2010). Departure of the most active members of the population may mean not only changes in the volume of food production but poorer access to food resources by older members of the household who have to shift their dependence to more distant relatives or neighbours.

To illustrate the above conceptualisations we report in the next section an example of some empirical work conducted in rural Bangladesh.

SOCIAL NETWORKS PROTECTING AGAINST ADVERSITY

In 2004 Kaberi Gayen completed research on the association of social networks to the uptake of contraception in rural Bangladesh (see Gayen 2004 and Gayen and Raeside 2010). This research involved interviewing 725 women of childbearing ages in seven non-contiguous rural villages. A structured interview was used based on a questionnaire designed to find out about conventional attributes such as women's socio economic backgrounds, how they received information and about women's social networks. This part involved women reporting on up to five other women who they were in contact with. The women were asked who they were in contact with, how frequently they were in contact and the importance of the contact. Frequency and importance were measured on one to five scales, where five was very frequently/very important. The product of frequency and importance was used to approximate the strength of ties between women. This study was relatively unique in that almost full networks were explored as every women in the villages were interviewed provided they consented, (less than five per cent did not wish to take part and also the boundary to the network was well defined in the sense that women tend not to travel, especially if not accompanied by their husband or other relation and the villages chosen were separated by several kilometres from other villages.) It is relevant to focus on women as they follow traditional lives being economically determined by their husband, and provide most of the caring duties and are likely to be the most affected and most vulnerable to the effects of climate change.

Here four of the villages used in Gayen's study will be re-examined focussing on the likelihood of a women and her family suffering food deficit, the villages were selected on the basis that the main form of production is agriculture. Food deficit will likely be one of the first symptoms of climate change. The four villages are located in the inland divisions of Rajshahi and Dhaka and in the coastal areas of the divisions of Barisal and Chittagong. The village in Rajshahi division has a history of suffering from severe drought while the village in Dhaka division has a history of flooding from the River Jamuna. The coastal divisions are vulnerable to cyclones, rising sea levels and storm surges – which will through changing salinity levels have particular detrimental effects on agriculture.

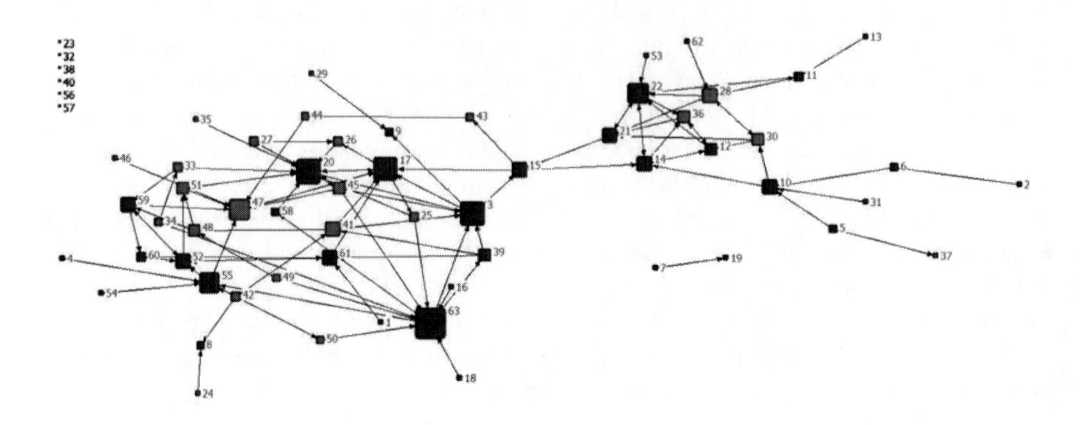

Figure 1. Sociogram of women's network of a village in Dhaka division.

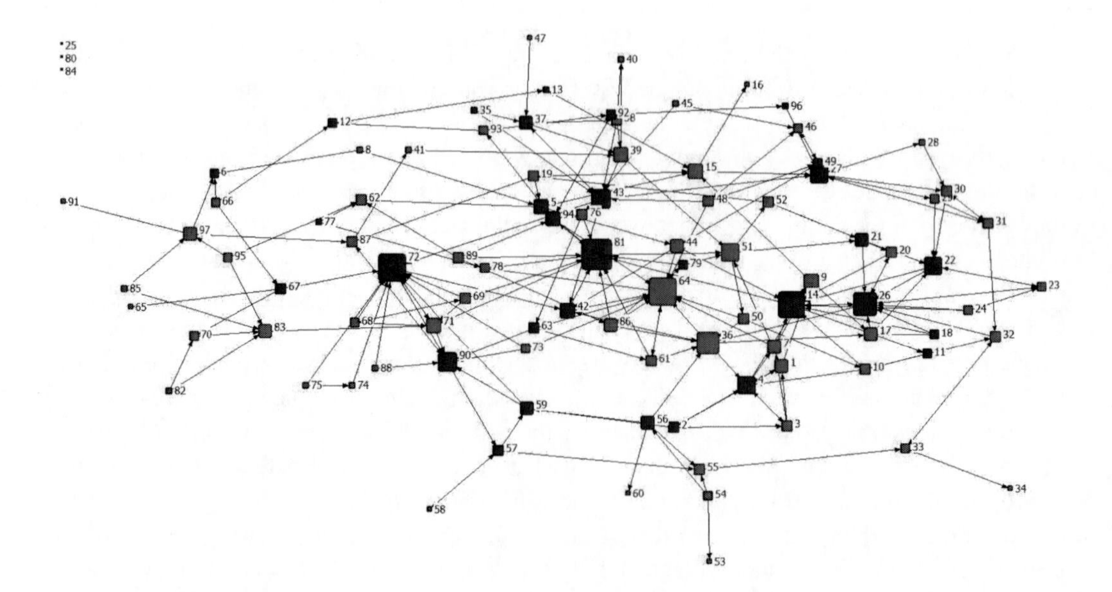

Figure 2. Sociogram of women's network of a village in Chittagong division.

To preserve anonymity the villages will be labelled A (located in Rajshahi division), B (located in Dhaka), C (located in Barisal division) and D (located in Chittagong division) and the names of the women in the network are replaced with numbers. The sample sizes were 63, 97, 98 and 80 for villages A, B, C and D respectively. The matrices of contacts were formed weighted by tie strength and entered in to the software UCINET 6 (Borgatti et al., 2002) to obtain via NETDRAW sociograms and measures of the network properties.

Examples of two of the sociograms are displayed in Figures 1 and 2. The nodes in these are different sizes to reflect the size of degree centrality (amount of connections weighted by strength).

The red nodes are women who report to be in food deficit. The majority of central actors are not in food deficit and it seems that those in food deficit seem to be connected to others reporting food deficit that is they are connected to others who by virtue of economic difficulty will have low social capital.

Yet again and more clearly this time those not reporting food deficit are more connected and tend to have higher centrality scores.

How measures of in and out degree are associated with different levels of food deficit is displayed in error bar plot shown Figure 3. This figure shows the 95% confidence interval around the mean of each centrality measure and it is apparent that higher levels of the centrality measures are associated with not being in food deficit.

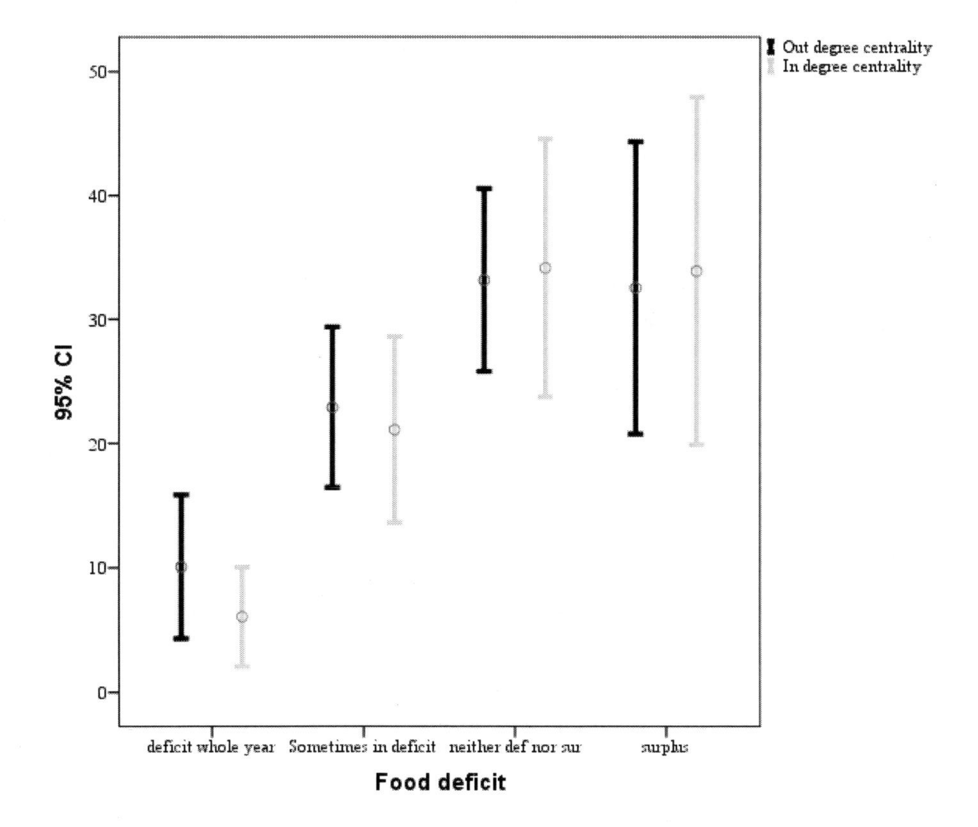

Figure 3. How centrality varies with level of being in food deficit.

The probability of being in food deficit is now related to women's centrality in the network while controlling for attribute based variables such as women's autonomy, family size and age. The variables in the model are described in Table 1 and their Pearson correlation coefficients to a four point scale of the degree of food deficit.

Table 1. Control Variables

Variable	Description	Summary
Age	Chronological age of women	Mean = 28 rears range = 15 to 50 years
Number of family members	Number of adult family members who live in the village	Mean 5.69 range 2 to 19
Number of children	Number of children ever born to the woman	Mean = 2.72 (range 0 to 9)
Female Autonomy Score	A variable formed from the first principle component of women's years of education, Likert scale measures of ability to travel unaccompanied and decision making capacity.	Range -2 to 1
Housing score	A standardised variable formed from Likert scale measures of the quality of the house floor, walls and roof. This is also a proxy measure of economic well-being	Range -2 to 2
Family member lives abroad	Binary variable 0 if no one and 1 is at least one family member overseas	Percentage of family members who live abroad 14%
Micro credit affiliation	Binary variable 0 if no one and 1 if affiliated to a micro credit organisation	Percentage who are affiliated 36%
Religion	1 if Hindu and 0 otherwise	Percentage Hindu 42.7%*

*In Bangladesh less than 10% of the population are Hindu; the reason for the proportion of Hindu's being so high is that village A is primarily Hindu.

A logistic regression model was constructed using the centrality variables and the control variables to predict the likelihood of being in food deficit (coded 0) or not in food deficit (coded 1). Dummy variables were used to represent the study areas to reflect local circumstances causing differences in variability between the regions. Wald's backwards elimination was used to exclude variables which were found not to be significant and to help avoid multicolinearity amongst the so called independent variables, for details, see Hosmer and Lemeshow (2000).

The model developed is presented in Table 2, this model showed adequate fit as the Hosmer and Lemeshow test for not being an adequate fit was not significant ($P = 0.622$) and the pseudo R^2 values of Cox and Snell and Nagelkerke were respectively 16.7% and 22.4% and correctly predicted almost 72% of those in food deficit.

Table 2. Logistic Regression model of the likelihood of being in food deficit

Variable	Coefficient	Standard Error	P value	Odds Ratio
In degree centrality	.014	.006	.034	1.014
Out degree centrality	.009	.004	.460	1.009
Number of children	-.313	.097	.001	.731
Family member lives abroad	.963	.308	.002	2.620
House hold score	.407	.151	.007	1.503
Study area (Baseline = Rajshahi)			.000	
Dhaka	.081	.361	.821	1.085
Barisal	-1.607	.406	.000	.200
Chittagong	-.977	.347	.005	.376
Constant	.716	.390	.066	2.046

From this model the effect of social networks (both measures of centrality in the networks) is protective in allowing those who are more connected to be resilient to food scarcity. Having family members living overseas also makes for greater resilience to food deficit. As one would expect those with higher economic assets, as measured by the housing score, have significantly greater resilience and those with more children tend to suffer significantly more from food shortages.

It is to be noted that the villages in the coastal divisions of Barisal and Chittagong are associated with more food deficit than the inland regions. The model however, leaves much unexplained; for instance only 62% of those not experiencing food deficit are correctly predicted. It is of note that micro credit affiliation often perceived as a panacea for those struggling to develop is not significant and neither is religion and number of family immediate members. It is perhaps comforting that the number of family members is not significant as consequences of population aging and migration away from rural areas that support from extended families is declining. There are many other variables that need to be considered such as the type of agriculture followed, interventions by government and NGOs, the local geography and environmental conditions.

Conclusion and Discussion

Populations are becoming older, meaning that there are proportionately fewer younger family members available to provide care and support. This ageing effect is exacerbated in many rural communities by migration of young workers to urban centres. A need is developing to rely on support from friends and the institutions provided by the state. However, in developing countries sufficient state support is unavailable for many. It has and will become important the networks one is connected to and how these can function to provide support. The other global force is climate change which is predicted to lead to unstable weather patterns leading to drought, flooding, rising sea levels and extreme storms. This is going to be very disruptive to those living in subsistence agriculture. Again the state or overseas aid cannot be expected and local community action is needed. To cope with these changes, social networks, if they function in a supportive way, can have a protective effect and facilitate resilience to these changes.

There is a growing body of literature, some of which is outlined in this chapter that illustrates the importance of social networks to promote wellbeing. The beneficial effects arise from connecting people together that will allow the provision of emotional support and physical help. Further by connecting to human and social capital outside the village gives important conduits to allow the diffusion of knowledge and modern ideas in to the community. This might be done via family members who have migrated to cities or overseas who provide bridges of knowledge flow in to the community. This will allow education to understand the consequences of climate change such as increased salinity levels and how to react to them by changing crop varieties and farming methods. This often challenges traditional ideas and so communication to facilitate ideational change is required. Gayen and Raeside (2006) and Kincaid (2000) show that this was achieved in promoting the norm of small family size being associated with improved economic wellbeing. Indeed Bangladesh acts as an exemplar for how if an idea can be accepted at community level then ideational shift can occur and for many this explains the dramatic drop in fertility levels from over seven children per family in the late 1970's to just over replacement levels in 2010.

The protective properties of social networks were illustrated in some rural villages in Bangladesh. These examples illustrated that being connected either as a source of information (in degree centrality) or as seeking information (out degree centrality) is associated with a lower likelihood of being in food deficit. Also important was having family members abroad which can be a source of the international transfer of ideas and this supports the findings of Scheffran et al., (2012) who illustrated from a study in Africa that "migrant social networks can help to build social capital to increase the social resilience in the communities of origin and trigger innovations across regions by the transfer of knowledge, technology, remittances and other resources".

Interestingly the studies reported on Bangladesh did not find that micro credit affiliation had a significant effect, a challenge to much of the development literature which argues of the benefits of micro credit organisations to allowing individuals to improve their lives (for example Mahmud 2003 and Amin et al 1998).

There is however a number of difficulties with the social network approach. Some of these emanate from the methodology in that the complete network is rarely known and there will always be unreported contacts, the nature of the associations is unclear and causality is a major concern, in the Bangladesh examples the fact that those who are more central tend to suffer less food deficit might be because not being in food deficit makes them popular. Also not all networks are protective and progressive, some networks might have people with very traditional ideas in central positions and this might act to inhibit the diffusion of new ideas; an example in relation to accessing health professionals is given by Gayen and Raeside (2007).

Nevertheless in nations such as Bangladesh especially when there is little alternative people need to work together to face the impending challenges and a way to achieve this is to promote effective and protective social networks. This fit well with Sen's (2000) capability approach that people should be empowered by developing their human and social capital, by developing their knowledge and skills and by linking individuals and networks to elites.

Thus the recommendations are to facilitate the development of local community networks. It cannot be assumed that these will just evolve. It should be a prerogative of a nation and states to develop the capabilities and the capacities of their populations. In Bangladesh it is time for a network of community workers who interact with local communities to facilitate the development of networks and these workers could also serve the role of bring knowledge to the villages and demonstrating and supplying information communication technology as a means of knowledge acquisition.

REFERENCES

Adams R. H. Jr. (2006). Remittances and poverty in Ghana. *World Bank Policy Research Paper 3838. Washington*, DC.

Adger, W. N., (2003). Social capital, collective action, and adaptation to climate change. *Economic Geography* 79 (4): 387–404.

Adger N., Paavola, J., Huq, S. and Mace, M. J. (2006). *Fairness in Adaptation to Climate Change. MIT* Press: Cambridge.

Afsana K., (2004). The tremendous cost of seeking hospital obstetric care in Bangladesh, *Reproductive Health Matters,* 12: 42-48.

Agrawala,S., Ota, T., Ahmed, A. U., Smith, J., and van Aalst, M., (2003). *Development and climate change in Bangladesh: Focus on coastal flooding and the Sundarbans,* OECD, France.

Ahmed S. M., Adams A., Chowdhury, M. and Bhuiya A., (2000).Gender socioeconomic development and health-seeking behaviour in Bangladesh, *Social Science and Medicine,* 51: 361-371.

Ali, A., (1999). Climate change impacts and adaptation assessment in Bangladesh, *Climate Change Research,* 12: 109 –116.

Amin, R., Becker, S. and Bayes, A. (1998). NGO-Promoted Microcredit Programs and Women's Empowerment in Rural Bangladesh: Quantitative and Qualitative Evidence. *Journal of Developing Areas* 1998, 221- 236.

Andrews, R. and Entwistle, T. (2010) 'Does Cross-Sectoral Partnership Deliver? An Empirical Exploration of Public Service Effectiveness, Efficiency and Equity', *Journal of Public Administration Research and Theory,* 20(3): 679-701.

Balkundi, P. and Kilduff, M. (2005). The ties that lead: A social network approach to leadership. *The Leadership Quarterly* 16: 941– 961.

Barrientos, A. Gorman, M. and Heslop, A., (2003). Old Age Poverty in Developing Countries: Contributions and Dependence in Later Life, *World Development,* 31, 3, 555-570.

Bebbington, A. (2004). Social capital and development studies 1: critique, debate, progress? *Progress in Development Studies,* 4: 343-349.

Bebbington, A., Guggenheim, S., Olson, E. and Woolcock, W., (2004). Exploring Social Capital Debates at the World Bank, *Journal of development Studies,* 40, (5): 33-64.

Bongaarts, J., and Watkins, S. C. (1996). Social interactions and contemporary fertility transitions. *Population & Development Studies,* 22, 639-682.

Borgatti, S.P., Everett, M.G. and Freeman, L.C. 2002. *Ucinet for Windows: Software for Social Network Analysis. Harvard*, MA: Analytic Technologies.

Cain, M. (1986). The consequences of reproductive failure: dependence, mobility, and mortality among the elderly of rural South Asia. *Population Studies* 40(3):375-88.

Choudhary, S. R., (2013), Socio-demographic Conditions of Rural Aged in Bangladesh, *Middle East Journal of Age and Ageing,* 10, 43-47.

Cleland, J., and Wilson, C. (1987). Demand theories of the fertility transition: an iconoclastic view. Population Studies, 41, 5e50.

Daraganova, G., Pattison, P., Koskinen, J., Mitchell, B., Bill, A.., Watts, M. and Baum, S., (2011)., Networks and geography: Modelling community network structures as the outcome of both spatial and network processes. Social Networks, 34(1): 6-17.

Das, J., Do, Q-T., Friedman, J., McKenzie, D. and Scott, K., (2007). Mental health and poverty in developing countries: Revisiting the relationship, Social Science and Medicine, 65, 467-480.

Datta, S.K. and Nugent, J. (1984). Are old-age security and utility of children in rural India really unimportant? Population Studies 38:507-509.

DFID, (2009). Water off a duck's back, Developments, 46, 12-13.

Doreian, P.and Conti, N., (2012). Social context, spatial structure and social network structure. Social Networks, 34(1): 32-46.

Falkingham, J., and Namazie, N., (2002). Measuring health and poverty: a review of approaches to identifying the poor, DFID Health Systems Resource Centre, London.

Farid, K. S., Ahmed, J. U., Sarma, P. K., and Begum, S., (2013). Population Dynamics in Bangladesh: Data sources, current facts and past trends, J. Bangladesh Agril. Univ. 9(1): 121–130.

Findlay, A., (2005). Vulnerable spatialities, Population, Space and Place, 11, 429-440.

Freeman, L. C., (2004). The Development off Social Network Analysis, Booksurge, LLC, North Charleston, South Carolina.

Friedman, L (2009). How Bangladesh Is Preparing for Climate Change, Scientific American, March.

Gayen, K. and Raeside, R. (2006). *Communication and Contraception,* Health Care Quarterly, 9(4): 110-22.

Gayen, K., and Raeside, R. (2006). Communication and contraception in rural Bangladesh. World Health & Population. *Also in Healthcare Quarterly* 9 (4): 110-122.

Gayen, K., and Raeside, R. (2007). Social networks, normative influence and health delivery in rural Bangladesh. *Social Science and Medicine,* 65(6): 900-914.

Gayen, K. (2004). Modelling the influence of communication on fertility behaviour of women in rural Bangladesh. *PhD thesis,* Edinburgh, UK: Centre for Mathematics and Statistics, Napier University.

Gayen, K. and Raeside, R. (2010a). Social networks and contraception practice of women in rural Bangladesh, *Social Science and Medicine,* 7(9): 1582-1592.

Gayen, K. and Raeside, R., (2010b). Communicative actions, women's degree of social connectedness and child mortality in rural Bangladesh. *Child: Care, Health and Development,* 36(6): 827–834.

Giddens A., (1984). *The Constitution of Society* Polity Press: Cambridge.

Granovetter, M. (1973). The strength of weak ties. *American Journal of Sociology,* 78, 1360-1380.

Gupta, S., Pattillino, C. A. and Wagh, S. (2009). Effect of Remittances on Poverty and Financial Development in Sub-Saharan Africa, *World Development,* 47, 1, 105-114.

Haines, A., Kovats, R. S., Campbell-Lendrum, D. and Corvalan, C., (2006). Climate change and human health: Impacts, vulnerability and public health, *Public Health,* 120, 585-596.

Harmeling, S. and Eckstein, D. (2012) *Global Climate Risk Index* 2013, Germanwatch e.V, Bonn.

HelpAge International Asia Pacific Regional Development Centre (2000). *Uncertainty rules our lives: the situation of older people in Bangladesh: Situation Report.* London: HelpAge International.

Hosmer, D. W. and Lemeshow, S., (2000). *Applied Logistic Regression,* wiley, Chichester.

Jesmin, S. S. and Ingman, S. R. (2011). Social Supports for older adults in Bangladesh, *Journal of Aging in Emerging Economies,* www.kent.edu/sociology/resources/ jaee/upload/jesmin_ingman.pdf.

Johnson, C. A. and Krishnamurthy, K., (2010). Dealing with displacement: Can ''social protection'' facilitate long-term adaptation to climate change? *Global Environmental Change,* 20(4): 648-655.

Kincaid, D. L. (2000). Social networks, ideation, and contraceptive behavior in Montgomery, M., and Casterline, J. (1996). Social learning, social influence, and new models of fertility. *Population and Development Review,* 22(Suppl.), 51e175.

Kniveton, D., Schmidt-Verkerk, K., smith, C. and Black, R., (2008). Climate change and migration: improving methodologies to estimate flows, *International Organization for Migration, Migration research Series 33,* Geneva, International Organization for Migration.

Knodel, J. and Debavalya, N. (1997). Living arrangements and support among the elderly in Southeast Asia: An introduction. *Asia-Pacific Population Journal* 12(4):5-16.

Knodel, J. and Ofstedal, M.B. (2002). Patterns and determinants of living arrangements. In. A.I. Hermalin (Ed.), *Well-being of the elderly in Asia: A four country comparative study* (pp. 143-184). Ann Arbor, MI: University of Michigan Press.

Leeson, G.W. (2006). My home is my castle-housing in old age: The Danish Longitudinal Future Study. *Journal of Housing for the Elderly* 20(3): 61-76.

Levkoff, S. E., McArthur, I. W., and Bucknall, J., (1995). Elderly mental health in the developing world, Social. *Science and Medicine,* 41, 7: 983-1003.

Licoppe, C. and Smoreda, Z., (2005). Are social networks technologically embedded?: How networks are changing today with changes in communication technology, *Social Networks,* 27, (4): 317-335.

Lloyd-Sherlock, P. (2010). *Population ageing and international development: From generalization to evidence,* Policy Press, Bristol.

Mahmud, S. (2003). *Actually how empowering is Microcredit? Development and Change,* 34(4): 577-605.

Mahmud, T., and Prowse, M., (2012). Corruption in cyclone preparedness and relief efforts in coastal Bangladesh: Lessons for climate adaptation?, *Global Environmental Change* 22, 933–943.

Matin, I. and Hulme, D., (2003). Programs for the poorest: Learning from the IGVGD Program in Bangladesh, *World Development,* 3, 647-665.

Mendelsohn, R., Dinar, A., Williams, L., (2006). The distributional impact of climate change on rich and poor countries. *Environment and Development Economics* 11, 159–178.

Montgomery, M. R., Kiros, G.-E., Agyeman, D. J., Casterline, B., Aglobitse, P., and Hewett, P. C. (2001). Social networks and contraceptive dynamics in Southern Ghana. *Working paper no. 153.* New York: Policy Research Division, Population Council.

Moreno, J. L. (1934). Who Shall Survive? *Nervous and Mental Disease Publishing Company,* Washington D. C.

Munsur, A.M., Tareque, I., and Rahman, M. (2010). Determinants of Living Arrangements, Health Status and Abuse among Elderly Women: A Study of Rural Naogaon District, Bangladesh. *Journal of International Women's Studies,* May 1. http://www.faqs.org/periodicals/201005/2093689401.html#ixzz10AuINQ1q

Nicholls, R. J., Wong, P. P., Burkett, V. R., Codignotto, J. O., Hay, J. E., McLean, R. F, Ragoonaden, S. and Woodroffe, C. D., (2007). Coastal systems and low lying areas in Parry, M. L ., Canziani, O. F., Palutikof, J. P., van der Linden, P. J. and Hanson, C. E., (eds), Climate Change 2007: Impacts, Adaptation and vulnerability. *Contributions of Working Group II to the Fourth Assessment Report of the Intergovernmental Panel on Climate Change,* Cambridge University Press, Cambridge, 315-356.

Nilsson, J., Grafstörm, M., Zaman, S. and Kabir, Z. N., (2005). Role and function: Aspects of quality of life of older people in rural Bangladesh, *Journal of Ageing Studies,* 19, 363-374.

Paul B. K. and Rumsey D. J. (2002). Utilization of health facilities and trained birth attendants for childbirth in rural Bangladesh: *An empirical study, social science and Medicine,* 54: 1755-1765.

Rahman, M. I., and Ali, A. M., (2009). *Population Aging and Its Implications in Bangladesh,* Jahangirnagar Review Part II Social Science,Vol.XXXI, http://papers.ssrn.com/sol3/papers.cfm?abstract_id=1349447.

Raeside, R., Gayen, K. and Canduela, J., (2009). Using Social Network Analysis to Study Social Capital and Development, *International Journal of Interdisciplinary Social Science,* 3, 10: 75-88.

Rindfuss, R. R., Jampaklay, A., Entwisle, B., Sawangdee, Y., Faust, K. and Prasartkul, P., (2004). *The Collection and Analysis of Social Network Data in Nang Rong, Thailand, in Martina Morris* (ed.) Network Epidemiology - A Handbook for Survey Design and Data Collection, Oxford, Oxford University Press

Rogers, E. M., (2003). *Diffusion of Innovation,* Free Press, CA.

Rogers, E. M., and Kincaid, D. L. (1981). *Communication networks towards a new paradigm for research.* New York: The Free Press.

Scheffran, J., Marmer, E. and Sow, P., (2012). Migration as a contribution to resilience and innovation in climate adaptation: Social networks and co-development in Northwest Africa, *Applied Geography* 33: 119-127.

Sen, A. K. (2000). *Development as Freedom,* Oxford University Press, Oxford.

Shahidul , M., Swapan, H., and Gavin, M., (2010). A desert in the delta: Participatory assessment of changing livelihoods induced by commercial shrimp farming in Southwest Bangladesh, *Ocean & Coastal Management* 54 (2011) 45-54.

Scott, J. P., (2000). *Social Network Analysis: A Handbook,* Sage, London.

Simmel, Georg (1955[1908]) *"The Web of Group Affiliations", in Conflict and the Web of Group Affiliations,* translated by R. Bendix. New York: Free Press

Sovacool, B, K., Anthony Louis D'Agostino, A. L., Rawlani, A., and Meenwat, H., (2012). Improving climate change adaptation in least developed Asia, *Environmental Science and Policy,* 2 1: 112-125.

Transparency International (2012). Corruption perception index 2012, http://cpi.transparency.org/cpi2012/results/, [accessed 2nd February 2013].

UNDP (2013). *Human Development Reports,* http://hdr.undp.org/en/reports/global/hdr2010/summary/poverty/, [Accessed February 2013].

US Census (2013). http://www.census.gov/population/international/data/idb/ information Gateway.php, [Accessed 12th February 2013].

Valente, T. W., (2010). Social Networks and Health, Oxford, Oxford University Press.

Verdery, A.M., Entwisle, B., Faust, K. and Rindfuss, R. R., (2012). Social and spatial networks: Kinship distance and dwelling unit proximity in rural Thailand. *Social Networks*, 34(1): 112-127.

Viry, G., (2012). Residential mobility and the spatial dispersion of personal networks: Effects on social support. *Social Networks,* 34: 59-72.

Wang, F., Y. and and Sun, Y.,(2006). Structure of peer-to-peer social networks, *Phys. Rev. E* 73, DOI. 10.1103/PhysRevE.73.0361.

Warner, K., Ehrhart, C., de Sherbinin, A., Adamo, S.B., Chai-Onn, T.C. (2009) In search of Shelter: Mapping the effects of climate change on human migration and displacement. A policy paper prepared for the 2009 Climate Negotiations. Bonn, Germany: United Nations University, CARE, and CIESIN-Columbia University and in close collaboration with the European Commission "Environmental Change and Forced Migration Scenarios Project", the UNHCR, and the World Bank.

Wenger C (ed) 1994 *Understanding Support Networks*, Avebury: Aldershot.

World Bank (2013). Bangladesh: Economics of Adaptation to Climate Change Study, http://climatechange.worldbank.org/content/bangladesh-economics-adaptation-climate-change-study, [Accessed 20th February 2013].

Zanoon, F., Findlay, A. and Lazaridis, G.(2006). *Old, poor and alone in Palestine,* Sociological Research Online, 11.4.

In: Social Networking
Editors: X. M. Tu, A. M. White and N. Lu
ISBN: 978-1-62808-529-7

Chapter 5

SOCIAL NETWORKING SERVICES AND ANALYSIS: THE THIRD REVOLUTION IN BEHAVIORAL RESEARCH?

Christopher M. Homan[1*] ***and Vincent M. B. Silenzio***[2†]
[1]Rochester Institute of Technology, Rochester, NY, US
[2]University of Rochester Medical Center, Rochester, NY, US

Abstract

We survey recent work that uses social networking services as a framework for social network analysis. Our survey is divided into three major categories of research design: observational, truly experimental, and quasi-experimental. Research in each of these categories addresses the fundamental problem of how to obtain a representative sample of a social network, among a host of other issues. Within each category, however, there are crucial differences in the technologies available and the problems addressed.

Keywords: Methodology, social network analysis, social networking services, literature review

1. Introduction

Emerging media have sometimes enabled dramatic increases in the power of behavioral research. Observes Reips (Rei13):

> Conducting studies via the Internet is considered a second revolution in behavioral research, after the computer revolution in the late 1960s and early 1970s that brought about many advantages over widely used paper-and-pencil procedures (e.g., automated processes, heightened precision).

*E-mail address: cmh@cs.rit.edu
†E-mail address: vincent_silenzio@urmc.rochester.edu

		Social presence/ Media richness		
		Low	Medium	High
Self-presentation/ Self-disclosure	High	Blogs	Social networking sites (e.g., Facebook)	Virtual social worlds (e.g., Second Life)
	Low	Collaborative projects (e.g., Wikipedia)	Content communities (e.g., YouTube)	Virtual game worlds (e.g., World of Warcraft)

Figure 1. Six categories of social media, from (KH10). The present review focuses on the upper middle quadrant (social networking sites) and experimental designs with the same social media profile.

With the emergence of online social media, we see a third revolution taking place, one that allows researchers to study human interaction directly—"in the field," so to speak—at scales and levels of time-and-space granularity previously unnattainable, and to translate these observational methods into quasi- and fully- experimental designs. This, in turn, has led to better opportunities for studying such basic questions as the role of selection vs influence, the social dynamics of infectious diseases, diffusion of innovations, the nature and purpose of social behaviors, and the role of social environment in mental health or healthful behavior.

This review focuses on a major part of the social media landscape: social networking services (SNSs) and the role they play in social network analysis (SNA) (WF94; CSW05), from the research design perspective. We survey recent research, organized around three major categories of designs (see http://socialresearchmethods.net/kb/destypes.php): randomized (true) experiments, quasi-experiments, and non-experiments (observational studies), and discuss some of the tradeoffs between each approach.

The spectrum of research that uses SNSs to collect data or conduct experiments is wide. It attracts investigators from sociology, economics, medicine, social psychology, statistics, computer science, physics, and informatics, among other fields. The examples from which we draw are biased toward behavioral health issues, because that is our main area of interest. A more comprehensive overview of the social or behavioral domains to which SNS-based designs apply is beyond the scope of this review.

1.1. Social Networking Services

Kaplan and Haenlein (KH10) develop a classification scheme for social media based on two dimensions: media richness/presence (SWC76; DL86) and self-presentation/disclosure (Gof59; SG03). See Figure 1. Their scheme leads to six main categories of social media, one of which are the social networking sites.

These sites have a number of characteristic properties beyond the two dimensions of Kaplan and Haenlein's model. See (boy09) for a detailed discussion. Three that are most significant to social network analysis are: (1) a sense of personal identity within the site, (2) some explicit notion of social space and locality, i.e., for each personal identity a friend or follower list, and (3) a computer-mediated channel for interacting with one's onsite social neighborhood. In essence, SNSs define explicitly a (computer-mediated) social network. Having access to this information can make many basic problems in social network analysis, such as helping participants recall the members of their social network, much easier to solve. Certainly, not all problems go away. For instance, a user's friends list may include people who have no functional relationship with the user.

"Social networking service" is a slightly broader term than "social networking site," as computer-mediated social networking may not occur on a website. Many of the fully- and quasi- experimental studies we survey here are in fact not web-based, are usually organized around a specific research goal, and may last only as long as it takes an experiment to run, which might be as brief as an hour. It is important to include such "*ad hoc*" SNSs in this review because, as we shall see, it is often difficult or infeasible to run experiments "in the field" on social networking sites. In such cases it becomes necessary to use *ad hoc* SNSs to recreate and isolate certain aspects of social dynamics, using the same basic interaction paradigms as their naturally occurring cousins.

1.2. Related Research

SNS-based research is part of the broader category of Web-based research (Rei13; Rei12). Reips (Rei05), citing (Bir04; KD00; Rei00; Rei97; Sch97) notes that while Web-based research often suffers from two major limitations: lack of control and observation of conditions, it has three main advantages: (1) access to large numbers of participants, (2) access to heterogeneous samples, and (3) cost-effectiveness. Other advantages include (4) automation and (5) access to archival data (KOB^{+}04). There are a number of risks and complications associated with Web-based research. (Rei13) summarizes them as relating to: design, security, recruitment, sampling, self-selection, multiple submissions, reactance-free question design, response time measurement, dropout, error estimation, data handling, data quality, privacy, and trust. The website http://psych.hanover.edu/research/exponnet.html is a clearinghouse for Internet-based psychology research. (Rei12) has a brief discussion of SNSs as a subfield of Web-based research (which also presents a nice set of principles for online survey design).

There are numerous studies on the relationship between online social networks and the "larger social context" (FMM10; ESL07). These are beyond the scope of this review. Our focus is rather on the use of social media to understand fundamental social phenomena, i.e., those that are not specific to one social milieu, but are more or less universal. The line between the two is admittedly hazy and we hope the reader will forgive some seemingly arbitrary omissions.

For more information about the implementation of social network analysis on SNSs, see the books (Rus11; Gol13).

2. Observational Designs

Observational studies are those in which the investigator cannot control the assignment of participants to treated and untreated groups. In SNS-based research subjects are often observed without their active participation and sometimes without even knowing that they are being observed. Such studies may make use of APIs provided by sites such as Facebook or Twitter (or by directly scraping the HTML) to extract user data such as: name, age, sex, sexual orientation, marital status, any text or media the users posts, when the post took place, and where the user was when he or she posted the information. One may also see the names or handles of some or all of the user's social connections (as with Facebook, MySpace or LinkedIn, where network connections are strictly reciprocal) or who the user follows (as with Twitter or LiveJournal, where network connections are directional, though of course may be reciprocated).

In some cases, researchers, working with an SNS provider, have obtained anonymized snapsots of an SNSs entire friendship graph. For example, Ahn et al. (AHK$^+$07) obtained two snapshots of the Cyworld (at the time the largest SNS in South Korea) network. The first snapshot, from November 2005, had 12 million users and 191 million friendship relationships and the second one, taken a year later, had 15 millions users and 291 million relationships. They discover that the degree distributions of these networks does not follow a single power law, as is true of most natural social networks, but rather seems to have two distinct scaling regions. They measure the assortativity (r) of these graphs, defined as

$$r = \frac{\langle k_i k_j \rangle - \langle k_i \rangle^2}{\sigma_k^2}, \tag{1}$$

where k_i and k_j are the degrees of nodes incident to edge ij, $\langle \cdot \rangle$ is the expected value function and σ_k^2 is the variance of the degrees (New02). Assortativity thus measures roughly how much any two linked nodes tend to have similar degree (in fact it is the Pearson correlation coefficient over all network dyads, hence the "r" notation). In natural networks, this number tends to be positive. In the Cyworld graph, assortativity is negative. They also study the effective diameter (i.e., the 90th percentile of the path length distribution) and the evolution of the clustering coefficient of this network and discuss its relationship to assortative mixing.

2.1. Sampling on Social Networking Services

In most cases it is difficult to obtain such complete information about an SNS. Thus, a significant body of work has developed that uses the explicit friendship structure SNSs provide to "crawl" the network in some fashion in order to get a representative sample of its population, usually without the active consent or even knowledge of the study participants. This practice has, over time—and as SNSs have upgraded their privacy policies and users have become more protective of their personal privacy—become increasingly difficult to execute. Crawls of this sort are usually breadth-first, i.e. starting from a "seed" participant(s), then sampling all the seed's friends, then all the seed's friends' friends etc. Breadth-first crawls have the advantage of capturing all relationships among a core group of sampled respondents, something that for certain kinds of social network analysis is critical to do. The downside to this approach is that it often biases the population-level statistics of the sample toward high-degree actors (KMT10; KMT11). This tradeoff between sample designs that accurately reflect the population statistics and those that capture the social dynamics of the network seems to be a fundamental problem.

Ahn et al. (AHK+07) conduct such breadth-first crawls on 100,000 each of MySpace and Orkut users and on 40,000 (via the friendship network) and 100,000 (via the testimonial network) Cyworld users. They study the degree distribution, clustering coefficient, assortativity, and effective diameter of these samples and of the two snapshots of the Cyworld graph they obtained (see above).

Mislove et al. (MMG+07) perform exhaustive bread-first searches (i.e., they only stop when they cannot find new, unvisited users) of Flickr, LiveJournal, Orkut, and YouTube. They discover that the amount of the network covered by this process varies dramatically, depending on the network (26.9% of the nodes were visited on Flickr, 95.4% on LiveJournal, 11.3% on Orkut, and YouTube coverage could not be estimated). Of these networks, all but Orkut have directed links. Flickr links can only be crawled in one direction, while LiveJournal can be crawled in both directions, and this may explain the discrepency in the coverage performance between those two.

They also compare a number of statistics taken from each of these samples, including link symmetry, degree distribution, path length, the "scale-free metric" (LADW05), assortativity, and clustering coefficient. They study the effect of removing increasing numbers of the highest degree nodes from the network. The latter method, inspired by (BKM+00), is intended to help understand the degree to which a small, highly connected core of users control the network. Among their most interesting results: they show that each of the three directed networks (Flickr, LiveJournal, and YouTube) have a high degree of link reciprocity, which is consistent with natural networks. The same three networks have degree distributions that fit power law models well. Orkut is the outlier with respect to degree distribution; the authors raise as possible explanations the poor coverage of the Orkut search sample and the fact that (as with LiveJournal) there are artificial limits placed on the number of links a user is able to have. With respect to clustering coefficient, they notice that, in general, lower degree nodes have a higher clustering coefficient in all four networks.

Mislove et al. (MKG+08) study the growth dynamics of Flickr. They crawl the network once per day, over a three-month period, observing 950,143 new users join and 9.7 million links form. They observe that 62% of all links were reciprocated and that 83% of all reciprocal link creations happened within 48 hours of the initial link creation. They show that users tend to create (respectively, receive) new links at a faster rate as their out-(respectively, in-) degree increases, and note that this is in accordance with the principles behind the Barabási-Albert (BA) preferential attachment model. However, unlike that model, in which links are chosen globally among all users, 50% of all new links are between users that were already connected (in their sample) and, of those, 80% are between users that were two hops apart, i.e., were friends-of-a-friend.

Wilson et al. (WBS+09) study the question of whether SNS friend lists are valid indicators of user interaction. They crawl selected Facebook regional networks (note: regional networks were removed from Facebook in 2009), and reach a sample of more than ten million users (940 million social links), and then measure their "interactions"—wall posts and photo comments—for a total of 24 million interaction events. They show, overall, that Facebook users interact with only about 60% of their Facebook friends, that the top 1% most active users account for 20% of all wall posts and 40% of all photo comments, and that the users with the 10% highest number of friends account for half of all interactions. They also study how interaction rates evolve over the lifetime of users and observe two trends: that new users (average Facebook lifetime of three weeks) tend to show a steep dropoff in activity soon after joining, while longer-time users (average lifetime of 20 months), due possibly to a steady increase in the number of friends with whom to interact, tend to increase their level of activity over time.

They introduce the concept of an interaction graph; a network parameterized by a time interval t and a minimum number of interactions n from one user to another for there to be an edge between them. They then compare these graphs, for various n and t, to their original friendship graph sample. They show, for instance, that the degree distributions of the interaction graphs trend with that of the friendship graph, though they do not scale equally.

One of the most potentially useful aspects of SNS-based research is in reaching "hidden" or difficult-to-reach individuals or populations. Early work by Silenzio et al. (SDT^{+}09) demonstrates that previously unimaginably large samples of lesbian, gay, and bisexual (LGB) youth can be reached via online social media. This work is motivated by the need to address the increased risk of suicide among young LGB individuals. Using automated, peer-driven methods, the researchers are able to easily identify a core sample of 100,014 LGB individuals between 16 and 24 years of age. The mean number of LGB peer "friends" in this sample is 137.5. They perform Monte Carlo simulations of peer-driven recruitment, which predict that sample sizes of up to 18,409 LGB individuals can be generated. The researchers thus show that reaching very large numbers of a previously difficult-to-reach population is now feasible.

2.2. Modeling Growth and Evolution

Kumar, Novak, and Tompkins (KNT10) study the *timegraphs* (the social graph as it evolves over time) of Flickr (from its birth) and Yahoo! 360 for a period of approximately 100 and 40 weeks, respectively. They notice three interesting trends in both graphs: (1) The edge density (fraction of actual edges compared to the total possible number of edges) evolve in three phases: initial dramatic growth, followed by a decline in edge density, and finally a steady and sustained increase. (2) The graphs each have a giant component and many smaller components, as do many natural social networks. New individuals attach themselves to either the giant component or one of the smaller ones and smaller components sometimes attach themselves to the giant component, but almost never do two smaller components connect directly. (3) The giant component effective diameter grows to a certain size, then tapers off (this is perhaps related to observation (1)). They build a generative model that seems to explain the second trend.

Backstrom et al. (BHKL06) study the process of group formation on the LiveJournal SNS and the DBLP citation network. LiveJournal, like many SNSs, allows users to create special interest groups for like-minded users to connect. They study two snapshots, spanning roughly four months, of 875 LiveJournal communities, focusing on the "fringe" of each community—those who are not members of the community but who have at least one friend who is a member—and study the conditions under which users in the fringe join the community in question. They discover that users on the fringe whose member friends are themselves tighly connected are much more likely to join the community than those whose member-friends are less tightly connected. They also show that the rate at which a community grows depends jointly on the size of the community, the size of the fringe, and the ratio of the fringe size to the community size, but none of these factors alone predicts this growth.

2.3. Predicting Emotion on Social Networking Services

Tang et al. (TZS+12; ZTS+10) propose a factor-graph model for predicting emotion on social networks that is notable in that they test it on both an SNS—LiveJournal—and a cellular phone network. Their approach is to choose parameters $(\{\alpha_k\}, \{\beta_{ji}\}, \{\lambda_i\})$ via the Metropolis-Hastings algorithm (CG95) that maximize the following likelihood function.

$$p(Y|G^t) = \frac{1}{Z}\exp\{\sum_{v_i \in V}[\sum_{x_{ik}^t \in X} \alpha_k f_k(x_{ik}^t, y_i^t) \tag{2}$$

$$+ \sum_{v_j \in NB(v_i)} \sum_{(y_i^t, y_j^{t'}) \in Y^t} -\beta_{ji}(t-t')(y_i^t - y_j^{t'})^2 \tag{3}$$

$$+ \sum_{(y_i^t, y_i^{t'}) \in Y^t} -\lambda_i(t-t')(y_i^t - y_i^{t'})^2]\} \tag{4}$$

where Z is a normalization factor, V is the set of users, $NB(v_i)$ are the network neighbors of user i, y_i^t is a binary value representing the emotional state of user i at time t as "positive" or "negative," X is a set of factors (which depends on the social network used), x_{ik}^t represents a particular factor at a particular time for a particular user, and f_k is a function that maps factor k to a value. The factors in X include location information, in the case of the cellular network, and in both networks (using SMS—short message service—data in the cellular network) a massive list of emotionally-charged words. They collect data on the emotions of the participants using a tailor-made cellphone application or a feature of LiveJournal that lets users indicate their mood.

They evaluate their model in terms of precision, recall, and F1-measure against a number of baseline models, including a support vector machine (SVM, using only user attributes, i.e., no information about network neighbors), SVM-net (which also includes among the attributes network neighbor information), naive bayes with the same factors as the SVM model, and naive bayes with the same features as the SVM-net model. They consider separately the predicted positive, negative, neutral, and average mood. Results are mixed, though their model generally outperforms the baselines, especially with respect to F1. On the negative side, they observe that removing information about a users' friends' emotional states has little effect on the performance of their model, which suggests that either social context has little effect on an individual's mood, as they measure it, or that their model fails to capture the underlying social dynamics of mood.

Thelwall et al. (TBP+10) develop a novel sentiment analysis algorithm, called SentiStrength, that they train using human coders on a corpus of 2,600 MySpace comments. The algorithm classifies words as positive or negative on a scale of 2 to 5 and uses expectation maximization, spelling correction, a booster word list, a negative word list, and other features to attenuate the human training signal. It should be noted that here, while words are rated on a one dimensional emotional scale, a body of text will have distinct positive and negative dimensions.

2.4. Twitter

Twitter deserves special mention, as it is a "microblogging" SNS targeted especially toward mobile users. Its interactions, or "tweets," are limited to 140 characters, and in this way it resembles SMS on cellular networks. It grafts this SMS-based interaction model to the typical features of an SNS, such as public user profiles and friends (called "followers" in Twitter, as they are directional) lists. As a result, Twitter, like all SNSs, has an explicit social network to study. Additionally, it may provide geographical information about where a tweet was posted, if the user enables this feature. Thus researchers can use it to study jointly social and geographical dynamics. This in turn creates opportunities to combine Twitter data with other geographically keyed data, such as Google maps or places.

Another remarkable feature of Twitter is the *garden hose* API. It allows one to query Twitter for certain features and receive an open stream of arbitrarily chosen tweets matching the query. The relative ease by which large amounts of data can thus be obtained has made Twitter one of the most popular sources of online social data. It would be a monumental task to comprehensively list all research using the Twitter API, and we will not attempt such a feat here. Java et al. (JSFT07) provide a broad study of the Twitter social network (including degree distribution, clustering, geographical distribution) and discuss the problems of community detection and trend tracking. Gonçalves, Perra, and Alessandro (GPV11) show that Twitter user activity shows a Dunbar's-number-like dropoff. Recent work has demonstrated that Twitter data can be used to predict a variety of social phenomena, including movie box-office revenues (AH10), elections (TSSW10), and flu epidemics (LDBC10; SKS12b).

Twitter's openness is a bit deceiving. While the garden hose and the rest of API make getting certain slices of data rather easily, other sorts of access are restricted. There are somewhat complex rules on how many users one can "follow," and this makes obtaining data on a large number of specific users difficult, and so in particular makes studying network dynamics particularly difficult. Nonetheless, there are a number of interesting studies on the network properties and dynamics of Twitter, as we now discuss.

2.5. Social Influence on Twitter

A fair amount of research exists on the prediction of the spread of information on Twitter and the identification of influential users (BHMW11; KLPM10; WLJH10; CHBG10; ZHVGS10). Bakshy et al. (BHMW11) studies 87 million tweets collected over a two-month period that contain a bit.ly URL. They identifed 1.6 million users, each of whom "seeded" (introduced into the network) an average of 46.33 URLs. They study cascades of influence in terms of whether a user tweets a URL that a follower subsequently reposts (either by retweeting or posting the same URL in a different tweet). They show that the size distribution of cascades follows a power law distribution. They define the influence of a user to be the logarithm of the mean size of all cascades for which the user was the seed. They fit this to a regression tree model (OS84) that depends on their number of followers, tweets, and retweets; date user joined Twitter; and statistics on the past influence of the user and the user's followers. They show that the most significant factors in predicting a user's influence are the number followers the user has and the past influence statistics of those followers. They also study the role content has to play inducing large cascades, by having

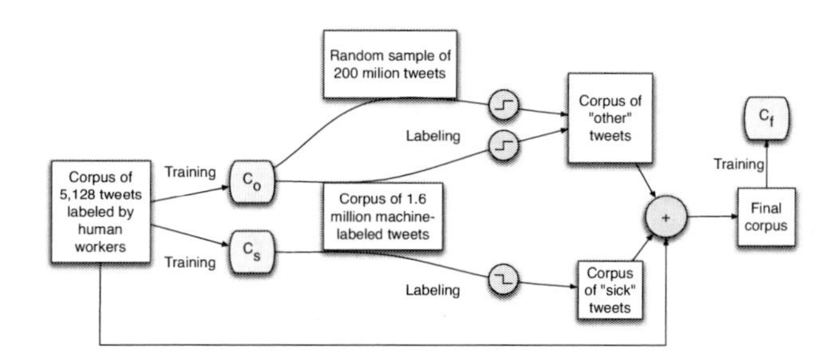

Figure 2. From (SKS12a). The SVM cascade model used to classify "sick" tweets.

Mechanical Turkers rate each URL, and noticed a tendency for more highly rated URLs to induce longer cascades.

Kwak et al. (KLPM10) crawl the entire space of Twitter users (41.7 million) and discover that the Twitter graph (1) has a degree distribution does not follow a power law and (2) has relatively low link reciprocity, both of which are different from most natural social networks. They note, however, that when this graph is limited to reciprocated relationships, it looks more like a natural network. They measure influence on Twitter using three metrics: number of followers, PageRank, and number of retweets, and show that each measure differs.

2.6. Predicting Disease on Twitter

Sadilek, Kautz, and Silenzio (SKS12a), building on prior work on disease surveillance on Twitter (PD12; LDBC10; Cul10; dQK10; CWZ08) use a cascading support-vector machine (CV95) to classify tweets as "sick" (i.e., mentioning flu-like symptoms) or other. They have multiple Amazon Mechanical Turk workers label each of 5,128 tweets as either "sick" or "other," and use this data to train two support vector machine (SVM) classifiers, one—C_s—tuned to sickness (i.e., heavily penalized for reporting an "other" tweet as "sick."), the other—C_o—tuned to other tweets. These classifiers are then used to train a third SVM on a set of 1.6 million tweets (both sets of tweets are from Paul and Dredze (PD12)), supplemented with an additional 200 million tweets C_o classified as "other." See Figure 2. The SVMs represent each tweet as a (sparse) vector of word uni-, bi-, and tri- grams. For instance, the tweet "I feel sick" would be represented as (i, feel, sick, i feel, feel sick, i feel sick). The feature vector space has more than 1.7 dimensions. Using this model on held-out training data, they achieve 0.98 precision and 0.97 recall.

They then study the effect of friendship and, taking advantage of Twitter's geolocation services, colocation on health tweets. They show that the likelihood of a user getting sick (i.e., sending a "sick" tweet) grows exponentially as the number of colocated sick Twitter users grows, and that this curve is dramatically steeper when those colocated users are also friends (i.e., mutually follow each other). They build on these observations in a second study (SKS12a), which uses a conditional random field model for predicting future illness. The model for each user consists of eight timeslices, each representing a day. Each timeslice t has a binary hidden node h_t, indicating sickness or health, and an observed node $\mathbf{o}_t$

containing a 25-element vector:

$$\mathbf{o}_t = (\text{weekday}, c_0, \ldots, c_7, u_0, \ldots, u_7, f_0, \ldots, f_7),$$

where c_i denotes the number of enounters with sick individuals i days ago, u_i denotes the number of unique sick individuals encountered, and f_i denotes the number of sick Twitter friends (all of which is determined by their SVM-based classification model). They perform a leave-one-out cross validation on 6,237 "geo-active" users (i.e., those Twitter users who enable automatic reporting of their location when they tweet) and obtain 0.94 precision and 0.18 recall.

Nagarahan, Purohit, and Sheth (NPS10) study a sample of nearly 1.7 million tweets on the Iran Election and the Health Care Reform Debate and extract from it the 250 most retweeted messages (accounting for 24,495 retweets). They notice two distinct patterns of content in these messages: (1) calls for action and (2) sharing of information. They define a *retweet network* as the graph induced by a particular retweeted message and observe that sparse retweet networks correlate with the presence of interpersonal pronouns (which are characteristic of calls for action) and verbs and dense networks correlate with the presence of details (characteristic of information sharing).

Huberman, Romero and Wu (HRW08) study a body of tweets from 309,740 Twitter users with an average of 255 tweets, 85 followers and 80 users each follows. They observed that the number of posts a users makes increases with the number of followers the person has, up to about 300 followers, and that a similar trend appears regarding number of friends (those who follow each other).

2.7. Studying Emotion on Twitter

A rather substantial body of work already exists on the use of Twitter to study emotion (BMZ11; DHK$^+$11; WCTS12; PGS12; KBO12; BGRM11; PGS12; BPM11; Moh12; GM11b; DCCG12; DCGC12; DCCH13a; DCC13; HAB$^+$12; TBP11; PP10), though comparatively little of it considers network properties. For instance, Golder and and Macy study aggregate global trends in "mood," and show, for example, that people wake up in a relatively good mood that decays as the day progresses (GM11b), Bollen et al. (BPM11) show that POMS-scored tweets are often tied to current events, such as elections and holidays.

Bollen et al. (BGRM11) define a friendship network over 4,844,430 Twitter users by including an edge exactly when a pair of Twitter users follows each other. Extracting the largest connected component from this graph yields a network of 102,009 Twitter users from which they estimate the subjective well-being (SWB) of 129 million tweets over six months. SWB is defined, using the OpinionFinder (OF) sentiment analyzer,[1] as

$$\mathcal{S} = \frac{N_p(u) - N_n(u)}{N_p(u) + N_n(u)}, \tag{5}$$

where u represents a user's tweets and N_p (respectively, N_n) is the number of tweets containing positive (respectively, negative) terms from the OF lexicon. They discover a bimodal distribution in the SWB score of their subjects. They measure the assortativity of

[1] http://www.cs.pitt.edu/mpqa/opinionfinderrelease/

the network with respect to SWB and observe a strong ($r = 0.443$), statistically significant ($p < 0.001$) score. They also define a notion of neighborhood assortativity, taken to be the Pearson correlation coefficient between an actor attribute and the mean value of that attribute over the actor's neighborhood. With respect to SWB, this has an even stronger, statistically significant score ($r = 0.689$).

Pfitzner, Garas, and Schweitzer (PGS12) study the factors involved in whether a tweet is "retweeted," i.e., "forwarded" in Twitter space. They use the SentiStrength classifier ($\text{TBP}^{+}10$) (see Section 2.3.) to rate tweets as positive or negative. They also introduce the notion of *emotional divergence*, taken as essentially the difference between the positive and negative ratings given by SentiStrength. They show that tweets with a high level of emotional divergence are more likely to be retweeted than those with a low level.

Kivran-Swaine and Naaman study the relationship between the tendency to express emotion and local network properties such as neighborhood size, the density of ego networks and reciprocity rate (KSN11). They distinguish between two types of tweets: *updates*, which are retweets or tweets that do not mention, or are not addressed to, other users, and *interactions*, which are addressed to specific followers. They handcode 194 tweets using the eight-primary-emotion scheme from Plutchik (Plu91) and then run a larger study, using the Linguistic Inquiry and Word Count (LIWC) text analysis program (http://www.liwc.net). They observe a strong positive correlation between interaction tweets expressing joy or sadness and the number of followers a user has, and a strong negative correlation between the same tweets and network density (significant correlations exist between emotional expression and reciprocity, but they are not as strong). Their results, they conclude, show a tendency for users to share more emotional information in larger, sparser networks.

3. Fully Experimental Designs

One of the fundamental problems in social network analysis is that it is difficult to separate network factors such as influence from homophily (ST11; Lyo11). Experimental designs using online social networks as the testbed can isolate and control for such factors. Such designs often require specialized software. See http://www.makeuseof.com/tag/the-5-best-open-source-social-networking-software/ for an overview of social-network-building tools.

3.1. Sampling on Social Networking Services

Gjorka et al. use a large number of repeated, randomized trials to study problems similar to those in Section 2.1. They study various methods for retrieving information from Facebook (GKBM10), including uniform (rejection) sampling (UNI) and multiple chain breadth-first search (BFS), random walks, reweighted random walks (RWRW), and Metropolis-Hasting random walks (MHRW) to estimate the degree distribution of the Facebook network. They also study the effectiveness of the Geweke and Gelman-Rubin diagnostics to determine sample size. Using UNI as ground truth, they showed that RWRW and MHRW were much more accurate than RW or BFS. RWRW was slightly more efficient (in terms of sample size needed) however, it requires that the sample be reweighted by each nodes neighborhood size, while the MHRW requires no such reweighting (it does however, require a more sophisticated sampling protocol).

3.2. Cooperative and Competitive Problem-Solving

There has been a fair amount of experimental work on how people use social networks to solve problems. Kearns and colleages (Kea12) studied a wide range of problems using specialized software that runs in a computer lab and provides an SNS-like experience for small communities of approximately 36 participants. Their experiments require the participants to solve problems having a game-theoretic flavor, using their SNS platform. Problems include: graph coloring (KSM06), coloring and consensus (JKV10), networked trade (JK08), networked bargaining (CJKT10), independent set (JKV11), biased voting (KJTW09), and network formation (KJV12). Subjects are randomly assigned positions in the network and different trials vary the network topologies or resources assigned to a given network position. Communication is limited to only only network neighbors and even then is highly restricted. For instance, in the coloring games participants might only see the colors that their network neighbors had chosen. Among their findings, the authors discover that when the networks are complex, having a global view of the network state (as opposed to just the view of local neighbor) leads to degraded performance.

Mason, Jones, and Goldstone (MJG08) design a computer-mediated social experiment to study the diffusion of innovation. They assign each member of a group ranging from 5 to 19 participants to a laboratory computer running a custom-made SNS that was connected via one of four patterns: lattice, fully connected, random, or small world. Participants were randomly assigned a position in the resulting network. They were given the task of guessing over the course of twenty rounds a number between 0 and 100. The number is assigned a payoff according to a fixed, continuous function that is not revealed to the participants. At the end of each round the choices and payoffs of each neighbor are revealed. Using strictly a bimodal (i.e., two local maxima) payoff function, they discovered that the fully connected graph led to the greatest amount of clustering in the final distribution of numbers selected. In a second study, they compared the speed it took the group to come within 0.5 standard deviations of the global maximum on one of two payoff functions, one unimodal, the other trimodal. They discover, overall, that convergence is much faster when the unimodal function is used, and that convergence was significantly slower on the lattice graph than on the fully-connected or small world graphs. In a final study, they used a highly eccentric, bimodal payoff function that "hid" the global maximum on a thin, steep hill. In this case, they found that the more spatially segregated networks, such as the lattice, had overall an easier time coming withing 0.5 standard deviations of the global maximum.

3.3. Cooperative Sense-Making

DiFonzio et al. (DBS$^+$13; SDYH10) built a web-based social network of sixteen individuals per trial for the purpose of studying a foundational meta-theory from the social psychology literature: dynamic social impact theory (DSIT) (LB). DSIT posits four patterns of social interaction: (1) attribute clustering, i.e., spatially-proximate pockets of attributes tend to emerge—that is, the attributes of actors sharing ties converge over time; (2) consolidation,

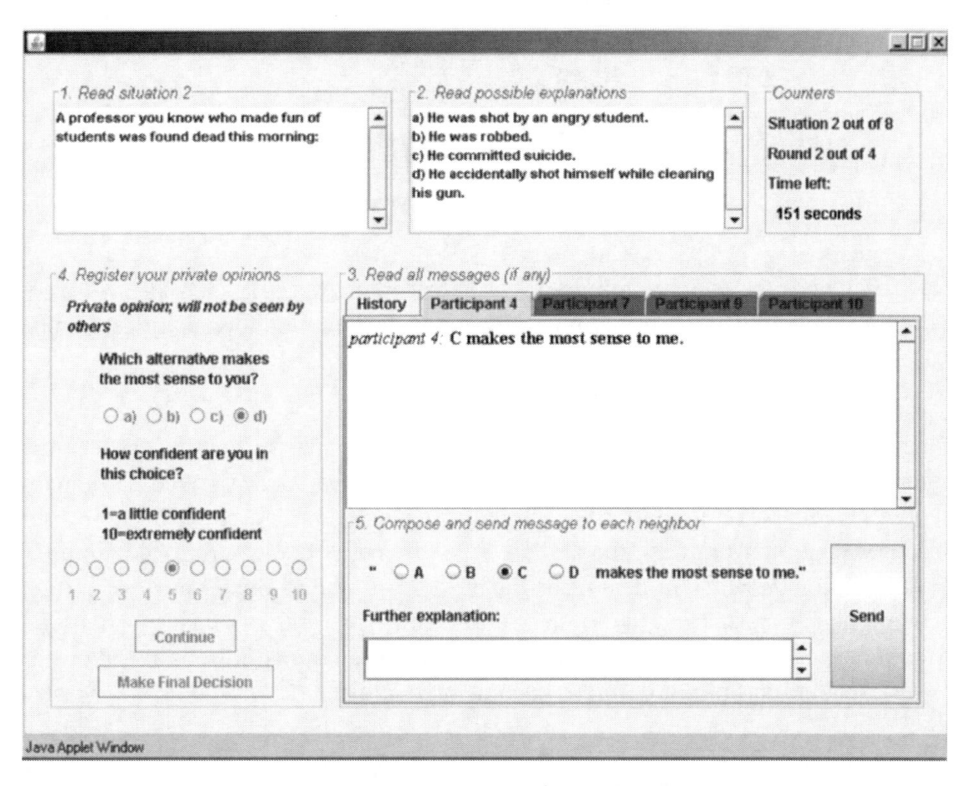

Figure 3. One of the SNS interfaces used in (DBS+13). It presents an ambiguous scenario in box 1, potential explanatory rumors in box 2, interactivity with network "friends" in box 3, and regular polling of individual opinions in box 4. In this way, interaction across the network can be tightly controlled and monitored as participants work to make sense of the scenario in terms of the rumors presented.

i.e., the overall diversity of attributes in the network tends to decrease over time (3) diversity, i.e., when initial diversity is of sufficient magnitude, minority attributes tend to persist, and (4) correlation, i.e., actors' attributes (or rather, pairs of attributes) tend to correlate over time, even on matters that have not been directly discussed.

The authors test the the first three of these patterns by looking at how network topologies affect the dynamics of rumors. From the social psychology standpoint, rumors are distinct from gossip, and serve as a form of sense-making about social problems involving a great deal of uncertainty, such as the death of a colleague or a new parking policy. Participants communicate electronically via an SNS interface regarding rumors that were presented to them. See Figure 3. Each study is parameterized by a set of scenarios, or topics for discussion, each with a selection or four possible explanations (i.e., rumors) for the scenario, and a set of virtual network types over which the subjects would discuss each scenario via their client computers. In particular, each participant was able to communicate directly with four (anonymized) others, depending on the underlying network, which shifted from time-to-time. Figure 4 shows the basic network topologies used. By anonymizing and randomly assigning individuals to positions in the networks, they were able to control for social selection biases.

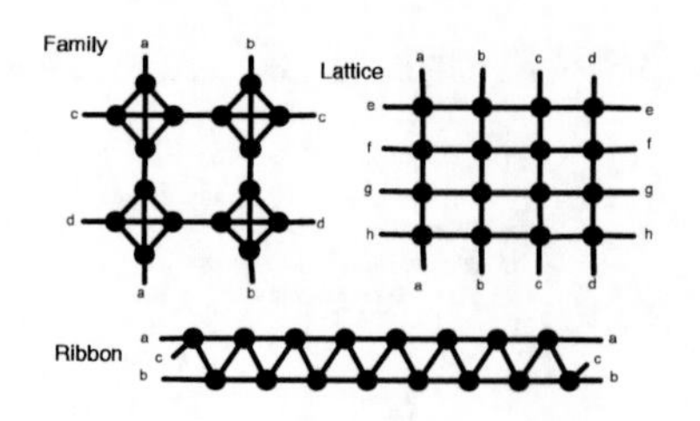

Figure 4. Three of the network topologies used by DiFonzio (DBS+13; SDYH10). These networks vary in terms of diameter and clustering. A fourth, random network structure was also used.

The particularly novel feature of their study and software was that it tightly controlled the number of interactions each participant was able to have, and it polled participants on their opinions of the rumors after each round of interactions, to see how their opinions evolved. Among their findings, the authors note that network clustering and an initial diversity of opinion on which rumors made the most sense lead to clustering of opinions (that is, network neighbors sharing the same opinions) on which rumors made the most sense. Also, confidence in a particular rumor increases with the amount of discussion allowed, and increases even more as rumor opinions become more clustered.

3.4. Health Behavior Dynamics

Centola (Cen10) studies the role social networks play in the adoption of new behavior. He hypothesizes that people are much more likely to adopt a new behavior when multiple friends first adopt it. He runs two studies to test this hypothesis on artificial social networks that ranged in size from 72 to 144 participants. In the first study, participants are recruited from health-interest World Wide Web sites and assigned to one of two networks, the first a regular lattice, the second a Watts-Strogatz small-world network (WS98). In either case, respondents interact though a web-based social networking service designed explicitly for the study. He then studies the spread of a health-related behavior (in this case, visiting a new health-related website) over a period of 21 days. His results show that more participants in the lattice network (which has a higher clustering coffecient than the small world network) adopt the behavior, compared to the small-world network, and also adopt the behavior more quickly. These results seem to support his hypothesis.

He also studies the role of homophily in social networks (Cen11). Once again, participants are assigned to one of two networks, though this time, they are recruited through an online fitness program. In this study, both networks are regular lattices with exactly the same network structure. In the first lattice users are assigned randomly, while in the second users are randomly assigned and then pairs of users are greedily swapped according to the pair which, if connected, would most increase overall network homophily with respect to age, sex, and obesity level. In this case, the health behavior studied was whether to use a web-based diet diary. Overall, the level of adoption was much lower than in the first study, but his results are nonetheless significant and show that, overall, adoption is higher in the homophilous network than in the "unstructured" one.

4. Quasi-Experimental Designs

Quasi-experimental designs are possible using data provided by SNS users. The researcher may be able to control the assignment of subjects using some criterion other than random assignment, such as using an eligibility cutoff mark. It is important to remain mindful that the lack of random assignment to experimental and control conditions can introduce challenges to internal validity within quasi-experimental designs. Research in this category tends to be a synthesis of the two previous categories and usually requires a dose of innovative social engineering and trial-and-error to achieve the right balance of virtues—naturalism and control, respectively—from the parent categories.

Much of the research in this category is in the form of simple surveys embedded within social networks (BKM11; LSHS13). However, even there, OSNs can provide key social context data, such as who the participant's network friends are, that is otherwise hard to obtain (Mar05). Facebook allows researchers to build applications that are embedded in the OSN itself and can thus be used to conduct a variety of research activities (QLCP+12; QL+12; BKM11; HTF08; GDM11)

Quercia et al. created a Facebook application that administers a personality test based on the five-factor model (CM92; GJE+06) and compares the results to the number of Facebook friends each participant has (QL+12). From a sample of 172,952 users they discovered a small ($R^2 = 0.064$) correlation between extroversion and number of friends (though the relationship between the logarithm of the average degree and extroversion, when graphed, is quite striking). They also test a prediction by Jaron Lanier (Lan10): that the most popular Facebook users will be those who are the best self-monitors of their Facebook personalities. They administer a self-monitoring and a five-factor personality test to 2,165 Facebook users. They find a relatively strong correlation between self-monitoring and extroversion (r = 0.312) and that, when controlled for self-monitoring, a much stronger correlation exists between extroversion and number of friends (r=0.16) than when self-monitoring is not controlled for.

Aral and Walker (AW12) use a Facebook application for rating films on a set of 7,730 users to study their susceptibility to influence, using a hazard model that distinguishes spontaneous from influenced behavior. When a user adopts or uses the application, it sends out messages to random subsets of the user's Facebook neighborhood. They discover (among other things) that susceptibility to influence descreases with age while influence increases, men are more influential and susceptible to influence than women, single and married people are more influential than those in relationships or in the "it's complicated" state, those who are engaged or "it's complicated" are most susceptible to influence, people in the same age range have the most influence, people are seldom both highly influential and highly susceptible to influence, influentials cluster on the network, and that there are more people with high influence scores than high susceptible scores.

4.1. Sampling on Social Networking Services

As we mentioned in section 2.1., privacy concerns have made it increasingly difficult to sample SNSs by crawling them without active user participation. One approach that the present authors have investigated is the use of respondent-driven sampling (RDS) (Hec97;

Hec02) a method that thus far has mostly been used in live settings, and never to our knowledge in an SNS (though Gjoka et al. do use some of the bias estimators developed for RDS in their RWRW sampling process (GKBM10); see Section 3.1.). RDS is a chain-referral method, like breadth-first sampling. The main difference between the two is that participants in RDS studies are tasked to recruit new members into the study. Each respondent is given coupons that can then be passed to other members of the target population to redeem by joining the survey. The coupons are used to compensate respondents for performing recruitment, which helps reduce volunteerism bias. However, care must be taken to tune the compensation rate to the target audience. The number of coupons is limited to make the sampling process more like a random walk and less like an *ad hoc* convenience sample.

Online communities seem particularly sensitive to the compensation structure and other inducements to participate. For example, Cross et al. (SCC$^+$10) attempted to use peer-driven sampling approaches to help identify US Army personnel at risk for suicide. The rationale for using peer-driven approaches in this context was clear: Army personnel at increased risk for suicide are embedded in an institutional structure that does not always foster disclosure of symptoms, or which may unintentionally reduce the readiness of soldiers to seek help. Peer-driven sampling methods offer a potential way to circumvent normal communication channels in this setting, and allow the identifications of individuals at risk. Nonetheless, when the researchers attempted to recruit participants for the study, the typical growth dynamics observed in respondent driven sampling designs was not observed, and the recruitment phase of the study failed to move beyond the initial nine seed participants. None of the 23 peers invited by these seed participants responded, despite being offered financial incentives. In qualitative follow-up discussions with other Army personnel who were not involved in the study, the researchers found that this failure could be related to the unusually strong downward pressures on disclosure that exist in this population. They recommended that future respondent driven sampling designs account for such potential countervailing influences by adjusting the number of coupons made available to invite new participants, in order to preserve sample growth. On the other hand, Lytle et al. (LSHS13) attempted a survey of users of the SNS TrevorSpace and attracted so many seed users within a three-hour window that they spent their entire study budget on just the seeds.

The obvious disadvantage of RDS is that it is much slower and more expensive than the previous methods discussed here. As the number of peers a given respondent recruits is limited, it typically does not produce a complete picture of even a local region of the underlying network. Among its advantages are that RDS is designed to discover hidden communities, i.e., communities that lack standard sampling frames. Note that by this definition, nearly any community fits this description. In particular, it can be used to target specific in- and out-groups for quasi-experimental research. In the case of truly stigmatized communities, the fact that recruitment is tasked to members of the community helps them maintain anonymity. Another virtue is that it was designed as an asymptotically unbiased sampling method. This assertion has been strongly questioned (Wej09; GH10; TG11; GS10). Homan, Sell, and Silenzio (HSS13) conjecture that many of the features of SNSs, including an explicit friendship lists and efficient resampling, might in fact enable RDS to perform in a manner that is closer to the ideal settings in which asymptotically unbiased sampling is guaranteed.

4.2. Suicide on Mixi

Masuda, Kurahashi, and Onari (MKO12) compare 9,990 users of mixi belonging to one of seven mixi user groups on suicide (the test group) to a randomly-drawn group of 228,949 mixi users who belong to none of these groups and have at least two friends (the control group). They consider a number of independent variables within each group: age, gender, number of mixi communities a participant belongs to ("community number"), number of network friends (which are undirected in mixi), local clustering coefficient, homophily—defined here as the fraction of neighbors who belong to one of the seven mixi suicide groups—and registration date (i.e., when a user joined mixi). They observe significant differences in each independent variable between the control and test groups (both in terms of Student's t-test and the Kolmogorov-Smirnov test). From a univariate logistic regression perspective, the three most significant variables are commuity number (area under the curve, or AUC, 0.867), clustering coefficient (AUC 0.690), and homophily (AUC 0.643). The remaining variables have marginal AUC values. A multivariate logistic regression on all indepedendent variables yielded an AUC of 0.873, which suggests that the community number variable accounts for most of the variance.

4.3. Discovering Depressive Disorders on Twitter

De Choudhury and colleagues (DCGCH13) leverage continuing streams of evidence on posting activity from Twitter that can reflect the underlying psychological states of Twitter users and the social milieus into which these are embedded. Moreover, their work corroborates their earlier findings (DCGC12; DCCG12), and those of others (GM11a; DHK$^+$11), that it is possible to predict in an unobtrusive and fine-grained manner individuals' vulnerabilities to subsequently developing depression. Their recent work focuses specifically on depressive disorders (DCCH13b; DCGCH13). They collect labels to consider as ground truth regarding the presence of major depression using crowd sourcing. Crowdsourcing for natural language processing tasks can be an efficient, timely, and inexpensive means to rapidly access behavioral data from a diverse population (SOJN08). The researchers analyze a full year of year of tweets from before the onset of depression for those who screen positive for major depression, using standard screening instruments, or who reported a diagnosis at some point during the previous year. In addition, they collect a full year of tweets for non-depressed participants in the crowdsourcing activity. The researchers measure behavioral attributes relating to social engagement, emotion, linguistic style, ego network, and explicit mentions of antidepressant medications. The researchers are able to build a major depression classifier using measures for these attributes that can predict outcomes with 70% accuracy and a precision of 0.74 (DCGCH13).

"Depression," of course, is not a unified diagnosis. In addition to the quintessential category of Major Depressive Disorder, many other favors of affective disorders overlap significantly in terms of signs, symptoms, and clinical course (Ass00). Parsing these different diagnostic categories may require additional contextual information, for example gender. In related work, De Choudhury et al. (DCCH13a), examine linguistic and emotional correlates for postnatal changes of new mothers, in order to predict extreme postnatal behavioral changes such as Postpartum Depression, using prenatal observations of tweets. The researchers quantify postpartum changes in 376 mothers using the dimensions of social engagement, emotion, social network, and linguistic style. Using a training set of observations of these measures before and after childbirth, they forecast significant postpartum changes in mothers with an accuracy of 71% when using observations about their prenatal behavior, which can be improved to an accuracy of 80-83% when also including tweets from the first 2-3 weeks of the postnatal period (DCCH13a).

5. Conclusion

The amount of research using SNSs to conduct social network analysis is massive and rapidly growing. Consequently, by the time this review is published it will probably already be partially obsolete. However, the field of opportunities is equally vast and to the extent that it helps new researchers gain a foothold in this arena, this review will prove valuable.

The astute reader will notice an imbalance in the literature between the three classes of research designs presented here. It should be clear that quasi- and fully- experimental designs are much costlier and more fraught with logistical dangers, and this is almost certainly why the volume of work in these categories is smaller. Nonetheless, we believe that those willing to put up with the extra trouble may be rewarded with a very open field of problems to choose from. And it seems to us that there is always a need for researchers willing to do the heavy—if frustrating—lifting required to accomplish such vexxing work.

Certainly one of the advantages of seeing these three fundamental approaches side by side is to realize that they form a continuum: as one moves from observational, through quasi-experimental, to experimental, there is a decrease in the amount of naturalness and scale and an increase in internal validity and cost. In the future, and especially with the rise of the big data paradigm, which in part deals with integration of heterogeneous data, we expect successful research to combine these methods, using experimental methods, for example, to ground behavioral models that are then applied to the Twitter datastream for large-scale observational studies, which can then be used to refine the model for later experimentation, and so on.

It is sometimes said that performing social network analysis requires a collaboration between a systems scientist, who understands the complex statistical modeling needed, and a domain expert, who understands the behavioral domain of interest. If that is so, then we humbly suggest that SNS-based SNA research requires a third collaborator: a computer or information scientist who understands the methodological limitations and virtues.

References

[AH10] S. Asur and B.A. Huberman. Predicting the future with social media. In *Web Intelligence and Intelligent Agent Technology (WI-IAT), 2010 IEEE/WIC/ACM International Conference on*, volume 1, pages 492–499. IEEE, 2010.

[AHK+07] Yong-Yeol Ahn, Seungyeop Han, Haewoon Kwak, Sue Moon, and Hawoong Jeong. Analysis of topological characteristics of huge online social networking services. In *Proceedings of the 16th international conference on World Wide Web*, pages 835–844. ACM, 2007.

[Ass00] American Psychiatric Association. *Diagnostic and Statistical Manual of Mental Disorders: DSM-IV-TR :Text Revision*. AMERICAN PSYCHIATRIC PRESS Incorporated (DC), 2000.

[AW12] Sinan Aral and Dylan Walker. Identifying influential and susceptible members of social networks. *Science*, 337(6092):337–341, 2012.

[BGRM11] Johan Bollen, Bruno Gonçalves, Guangchen Ruan, and Huina Mao. Happiness is assortative in online social networks. *Artificial life*, 17(3):237–251, 2011.

[BHKL06] Lars Backstrom, Dan Huttenlocher, Jon Kleinberg, and Xiangyang Lan. Group formation in large social networks: membership, growth, and evolution. In *Proceedings of the 12th ACM SIGKDD international conference on Knowledge discovery and data mining*, pages 44–54. ACM, 2006.

[BHMW11] Eytan Bakshy, Jake M Hofman, Winter A Mason, and Duncan J Watts. Everyone's an influencer: quantifying influence on twitter. In *Proceedings of the fourth ACM international conference on Web search and data mining*, pages 65–74. ACM, 2011.

[Bir04] Michael H Birnbaum. Human research and data collection via the internet. *Annu. Rev. Psychol.*, 55:803–832, 2004.

[BKM$^+$00] Andrei Broder, Ravi Kumar, Farzin Maghoul, Prabhakar Raghavan, Sridhar Rajagopalan, Raymie Stata, Andrew Tomkins, and Janet Wiener. Graph structure in the web. *Computer networks*, 33(1):309–320, 2000.

[BKM11] Moira Burke, Robert Kraut, and Cameron Marlow. Social capital on facebook: Differentiating uses and users. In *Proceedings of the 2011 annual conference on Human factors in computing systems*, pages 571–580. ACM, 2011.

[BMZ11] Johan Bollen, Huina Mao, and Xiaojun Zeng. Twitter mood predicts the stock market. *Journal of Computational Science*, 2(1):1–8, 2011.

[boy09] danah boyd. Why youth (heart) social network sites: The role of networked publics in teenage social life. 2009.

[BPM11] Johan Bollen, Alberto Pepe, and Huina Mao. Modeling public mood and emotion: Twitter sentiment and socio-economic phenomena. In *Proceedings of the Fifth International AAAI Conference on Weblogs and Social Media*, pages 450–453, 2011.

[Cen10] Damon Centola. The spread of behavior in an online social network experiment. *science*, 329(5996):1194–1197, 2010.

[Cen11] Damon Centola. An experimental study of homophily in the adoption of health behavior. *Science*, 334(6060):1269–1272, 2011.

[CG95] Siddhartha Chib and Edward Greenberg. Understanding the metropolis-hastings algorithm. *The American Statistician*, 49(4):327–335, 1995.

[CHBG10] Meeyoung Cha, Hamed Haddadi, Fabricio Benevenuto, and Krishna P Gummadi. Measuring user influence in twitter: The million follower fallacy. In *4th international aaai conference on weblogs and social media (icwsm)*, volume 14, page 8, 2010.

[CJKT10] Tanmoy Chakraborty, Stephen Judd, Michael Kearns, and Jinsong Tan. A behavioral study of bargaining in social networks. In *Proceedings of the 11th ACM conference on Electronic commerce*, pages 243–252. ACM, 2010.

[CM92] Paul T Costa and Robert R McCrae. *Revised neo personality inventory (neo pi-r) and neo five-factor inventory (neo-ffi)*. Psychological Assessment Resources Odessa, FL, 1992.

[CSW05] Peter J Carrington, John Scott, and Stanley Wasserman. *Models and methods in social network analysis*. Cambridge university press, 2005.

[Cul10] Aron Culotta. Towards detecting influenza epidemics by analyzing twitter messages. In *Proceedings of the first workshop on social media analytics*, pages 115–122. ACM, 2010.

[CV95] C. Cortes and V. Vapnik. Support-vector networks. *Machine learning*, 20(3):273–297, 1995.

[CWZ08] Bo Cowgill, Justin Wolfers, and Eric Zitzewitz. Using prediction markets to track information flows: Evidence from google. *Dartmouth College*, 2008.

[DBS+13] Nicholas DiFonzo, Martin J Bourgeois, Jerry Suls, Christopher Homan, Noah Stupak, Bernard P Brooks, David S Ross, and Prashant Bordia. Rumor clustering, consensus, and polarization: Dynamic social impact and self-organization of hearsay. *Journal of Experimental Social Psychology*, 49(3):378–399, 2013.

[DCC13] Munmun De Choudhury and Scott Counts. Understanding affect in the workplace via social media. 2013.

[DCCG12] Munmun De Choudhury, Scott Counts, and Michael Gamon. Not all moods are created equal! exploring human emotional states in social media. In *Sixth International AAAI Conference on Weblogs and Social Media*, 2012.

[DCCH13a] Munmun De Choudhury, Scott Counts, and Eric Horvitz. Major life changes and behavioral markers in social media: Case of childbirth. *Proc. CSCW 2013*, 2013.

[DCCH13b] Munmun De Choudhury, Scott Counts, and Eric Horvitz. Social Media as a Measurement Tool of Depression in Populations . In *5th ACM International Conference on Web Science (WebSci 2013)*, 2013.

[DCGC12] Munmun De Choudhury, Michael Gamon, and Scott Counts. Happy, nervous or surprised? classification of human affective states in social media. In *Sixth International AAAI Conference on Weblogs and Social Media*, 2012.

[DCGCH13] Munmun De Choudhury, Michael Gamon, Scott Counts, and Eric Horvitz. Predicting Depression via Social Media . In *7th International AAAI Conference on Weblogs and Social Media (ICWSM 2013)*, 2013.

[DHK+11] Peter Sheridan Dodds, Kameron Decker Harris, Isabel M Kloumann, Catherine A Bliss, and Christopher M Danforth. Temporal patterns of happiness and information in a global social network: Hedonometrics and twitter. *PloS one*, 6(12):e26752, 2011.

[DL86] Richard L Daft and Robert H Lengel. Organizational information requirements, media richness and structural design. *Management science*, 32(5):554–571, 1986.

[dQK10] Ed de Quincey and Patty Kostkova. Early warning and outbreak detection using social networking websites: The potential of twitter. In *Electronic Healthcare*, pages 21–24. Springer, 2010.

[ESL07] Nicole B Ellison, Charles Steinfield, and Cliff Lampe. The benefits of facebook friends: social capital and college students use of online social network sites. *Journal of Computer-Mediated Communication*, 12(4):1143–1168, 2007.

[FMM10] Sarah Jean Fusco, Katina Michael, and MG Michael. Using a social informatics framework to study the effects of location-based social networking on relationships between people: A review of literature. In *Technology and Society (ISTAS), 2010 IEEE International Symposium on*, pages 157–171. IEEE, 2010.

[GDM11] Anatoliy Gruzd, Sophie Doiron, and Philip Mai. Is happiness contagious online? a case of twitter and the 2010 winter olympics. In *System Sciences (HICSS), 2011 44th Hawaii International Conference on*, pages 1–9. IEEE, 2011.

[GH10] K. Gile and M. Handcock. Respondent-driven sampling: an assessment of current methodology. *Sociological Methodology*, 40(1):285–327, August 2010.

[GJE+06] Lewis R Goldberg, John A Johnson, Herbert W Eber, Robert Hogan, Michael C Ashton, C Robert Cloninger, and Harrison G Gough. The international personality item pool and the future of public-domain personality measures. *Journal of Research in Personality*, 40(1):84–96, 2006.

[GKBM10] Minas Gjoka, Maciej Kurant, Carter T Butts, and Athina Markopoulou. Walking in facebook: A case study of unbiased sampling of osns. In *INFOCOM, 2010 Proceedings IEEE*, pages 1–9. IEEE, 2010.

[GM11a] S A Golder and M W Macy. Diurnal and Seasonal Mood Vary with Work, Sleep, and Daylength Across Diverse Cultures. *Science (New York, N.Y.)*, 333(6051):1878–1881, September 2011.

[GM11b] S.A. Golder and M.W. Macy. Diurnal and seasonal mood vary with work, sleep, and daylength across diverse cultures. *Science*, 333(6051):1878–1881, 2011.

[Gof59] Erving Goffman. The presentation of self in everyday life. 1959.

[Gol13] Jennifer Golbeck. *Analyzing the Social Web*. Morgan Kaufman, 2013.

[GPV11] Bruno Gonçalves, Nicola Perra, and Alessandro Vespignani. Modeling users' activity on twitter networks: Validation of dunbar's number. *PloS one*, 6(8):e22656, 2011.

[GS10] Sharad Goel and Matthew J Salganik. Assessing respondent-driven sampling. *Proceedings of the National Academy of Sciences of the United States of America*, 107(1515):6743–6747, 2010.

[HAB$^+$12] Aniko Hannak, Eric Anderson, Lisa Feldman Barrett, Sune Lehmann, Alan Mislove, and Mirek Riedewald. Tweetinin the rain: Exploring societal-scale effects of weather on mood. In *Proceedings of the 6th International AAAI Conference on Weblogs and Social Media (ICWSM12)Dublin2012*, 2012.

[Hec97] Douglas D Heckathorn. Respondent-driven sampling: a new approach to the study of hidden populations. *Social problems*, pages 174–199, 1997.

[Hec02] Douglas D Heckathorn. Respondent-driven sampling ii: deriving valid population estimates from chain-referral samples of hidden populations. *Social problems*, 49(1):11–34, 2002.

[HRW08] Bernardo Huberman, Daniel Romero, and Fang Wu. Social networks that matter: Twitter under the microscope. *Available at SSRN 1313405*, 2008.

[HSS13] Christopher M Homan, Vincent Silenzio, and Randall Sell. Respondent-driven sampling in online social networks. In *Social Computing, Behavioral-Cultural Modeling and Prediction*, pages 403–411. Springer, 2013.

[HTF08] Jeffrey T Hancock, Catalina L Toma, and Kate Fenner. I know something you don't: the use of asymmetric personal information for interpersonal advantage. In *Proceedings of the 2008 ACM conference on Computer supported cooperative work*, pages 413–416. ACM, 2008.

[JK08] J Stephen Judd and Michael Kearns. Behavioral experiments in networked trade. In *Proceedings of the 9th ACM conference on Electronic commerce*, pages 150–159. ACM, 2008.

[JKV10] Stephen Judd, Michael Kearns, and Yevgeniy Vorobeychik. Behavioral dynamics and influence in networked coloring and consensus. *Proceedings of the National Academy of Sciences*, 107(34):14978–14982, 2010.

[JKV11] Stephen Judd, Michael Kearns, and Yevgeniy Vorobeychik. Behavioral conflict and fairness in social networks. In *Internet and Network Economics*, pages 242–253. Springer, 2011.

[JSFT07] Akshay Java, Xiaodan Song, Tim Finin, and Belle Tseng. Why we twitter: understanding microblogging usage and communities. In *Proceedings of the 9th WebKDD and 1st SNA-KDD 2007 workshop on Web mining and social network analysis*, pages 56–65. ACM, 2007.

[KBO12] Suin Kim, J Bak, and Alice Oh. Do you feel what i feel? social aspects of emotions in twitter conversations. In *Proceedings of the AAAI International Conference on Weblogs and Social Media*, 2012.

[KD00] John H Krantz and Reeshad Dalal. Validity of web-based psychological research. 2000.

[Kea12] Michael Kearns. Experiments in social computation. *Communications of the ACM*, 55(10):56–67, 2012.

[KH10] Andreas M Kaplan and Michael Haenlein. Users of the world, unite! the challenges and opportunities of social media. *Business horizons*, 53(1):59–68, 2010.

[KJTW09] Michael Kearns, Stephen Judd, Jinsong Tan, and Jennifer Wortman. Behavioral experiments on biased voting in networks. *Proceedings of the National Academy of Sciences*, 106(5):1347–1352, 2009.

[KJV12] Michael Kearns, Stephen Judd, and Yevgeniy Vorobeychik. Behavioral experiments on a network formation game. In *Proceedings of the 13th ACM Conference on Electronic Commerce*, pages 690–704. ACM, 2012.

[KLPM10] Haewoon Kwak, Changhyun Lee, Hosung Park, and Sue Moon. What is twitter, a social network or a news media? In *Proceedings of the 19th international conference on World wide web*, pages 591–600. ACM, 2010.

[KMT10] Maciej Kurant, Athina Markopoulou, and Patrick Thiran. On the bias of bfs. *arXiv preprint arXiv:1004.1729*, 2010.

[KMT11] Maciej Kurant, Athina Markopoulou, and Patrick Thiran. Towards unbiased bfs sampling. *Selected Areas in Communications, IEEE Journal on*, 29(9):1799–1809, 2011.

[KNT10] Ravi Kumar, Jasmine Novak, and Andrew Tomkins. Structure and evolution of online social networks. In *Link Mining: Models, Algorithms, and Applications*, pages 337–357. Springer, 2010.

[KOB+04] Robert Kraut, Judith Olson, Mahzarin Banaji, Amy Bruckman, Jeffrey Cohen, and Mick Couper. Psychological research online. *American Psychologist*, 59(2):105–117, 2004.

[KSM06] Michael Kearns, Siddharth Suri, and Nick Montfort. An experimental study of the coloring problem on human subject networks. *Science*, 313(5788):824–827, 2006.

[KSN11] Funda Kivran-Swaine and Mor Naaman. Network properties and social sharing of emotions in social awareness streams. In *Proceedings of the ACM 2011 conference on Computer supported cooperative work*, pages 379–382. ACM, 2011.

[LADW05] Lun Li, David Alderson, John C Doyle, and Walter Willinger. Towards a theory of scale-free graphs: Definition, properties, and implications. *Internet Mathematics*, 2(4):431–523, 2005.

[Lan10] Jaron Lanier. *You are not a gadget*. Vintage, 2010.

[LB] Bibb Latané and Martin J Bourgeois. Dynamic social impact and the consolidation, clustering, correlation, and continuing diversity of culture. *Blackwell handbook of social psychology: Group processes*, pages 235–258.

[LDBC10] V. Lampos, T. De Bie, and N. Cristianini. Flu detector-tracking epidemics on Twitter. *Machine Learning and Knowledge Discovery in Databases*, pages 599–602, 2010.

[LSHS13] Megan Lytle, Vincent M. B. Silenzio, Christopher M. Homan, and Pheonix Scheider. Suicidal and help-seeking behaviors among an online lgbt social network. Submitted, 2013.

[Lyo11] Russell Lyons. The spread of evidence-poor medicine via flawed social-network analysis. *Statistics, Politics, and Policy*, 2(1), 2011.

[Mar05] Peter V Marsden. Recent developments in network measurement. *Models and methods in social network analysis*, 8:30, 2005.

[MJG08] Winter A Mason, Andy Jones, and Robert L Goldstone. Propagation of innovations in networked groups. *Journal of Experimental Psychology: General*, 137(3):422, 2008.

[MKG+08] Alan Mislove, Hema Swetha Koppula, Krishna P Gummadi, Peter Druschel, and Bobby Bhattacharjee. Growth of the flickr social network. In *Proceedings of the first workshop on Online social networks*, pages 25–30. ACM, 2008.

[MKO12] Naoki Masuda, Issei Kurahashi, and Hiroko Onari. Suicide ideation of individuals in online social networks. *arXiv preprint arXiv:1207.0561*, 2012.

[MMG+07] Alan Mislove, Massimiliano Marcon, Krishna P Gummadi, Peter Druschel, and Bobby Bhattacharjee. Measurement and analysis of online social networks. In *Proceedings of the 7th ACM SIGCOMM conference on Internet measurement*, pages 29–42. ACM, 2007.

[Moh12] Saif M Mohammad. # emotional tweets. In *Proceedings of the First Joint Conference on Lexical and Computational Semantics-Volume 1: Proceedings of the main conference and the shared task, and Volume 2: Proceedings of the Sixth International Workshop on Semantic Evaluation*, pages 246–255. Association for Computational Linguistics, 2012.

[New02] Mark EJ Newman. Assortative mixing in networks. *Physical review letters*, 89(20):208701, 2002.

[NPS10] Meenakshi Nagarajan, Hemant Purohit, and Amit Sheth. The role of content on observed information diffusion in Twitter. *Proceedings of the WebSci10: Extending the Frontiers of Society On-Line,*, 2010.

[OS84] L Breiman JH Friedman RA Olshen and Charles J Stone. Classification and regression trees. *Wadsworth International Group*, 1984.

[PD12] Michael J Paul and Mark Dredze. A model for mining public health topics from twitter. *HEALTH*, 11:16–6, 2012.

[PGS12] René Pfitzner, Antonios Garas, and Frank Schweitzer. Emotional divergence influences information spreading in twitter. *AAAI ICWSM*, 2012:2–5, 2012.

[Plu91] Robert Plutchik. *The emotions*. University Press of Amer, 1991.

[PP10] Alexander Pak and Patrick Paroubek. Twitter as a corpus for sentiment analysis and opinion mining. In *Proceedings of LREC*, volume 2010, 2010.

[QL+12] Daniele Quercia, Renaud Lambiotte, , David Stillwell, Michael Kosinski, and Jon Crowcroft. The personality of popular facebook users. In *Proceedings of the ACM 2012 conference on Computer Supported Cooperative Work*, pages 955–964, 2012.

[QLCP+12] Daniele Quercia, Diego Las Casas, Joao Paulo Pesce, David Stillwell, Michal Kosinski, Virgilio Almeida, and Jon Crowcroft. Facebook and privacy: The balancing act of personality, gender, and relationship currency. In *Sixth International AAAI Conference on Weblogs and Social Media*, 2012.

[Rei97] Ulf-Dietrich Reips. Das psychologische experimentieren im internet. *Internet für Psychologen*, 2:319–344, 1997.

[Rei00] Ulf-Dietrich Reips. The web experiment method: Advantages, disadvantages, and solutions. *Psychological experiments on the Internet*, pages 89–117, 2000.

[Rei05] Ulf-Dietrich Reips. Web-Based Methods. In Michael Eid and Ed Diener, editors, *Handbook of Multimethod Measurement in Psychology*. American Psychological Association, Washington, DC, 2005.

[Rei12] Ulf-Dietrich Reips. Using the internet to collect data. 2012.

[Rei13] Ulf-Dietrich Reips. Internet-based studies. In Marc D Gellman and J Rick Turner, editors, *Encyclopedia of Behavioral Medicine*. Springer, 2013.

[Rus11] Matthew Russell. *Mining the Social Web: Analyzing Data from Facebook, Twitter, LinkedIn, and Other Social Media Sites*. O'Reilly Media, 2011.

[SCC+10] V.M.B. Silenzio, W. Cross, E. Caine, T. Prosser, and J. Cartwright. Freds: A novel adaptation of peer-driven sampling methods in online social media. In *Society for Prevention Research Annual Conference 2010*, June, 2010 2010.

[Sch97] William C Schmidt. World-wide web survey research: Benefits, potential problems, and solutions. *Behavior Research Methods, Instruments, & Computers*, 29(2):274–279, 1997.

[SDT+09] Vincent M B Silenzio, Paul R Duberstein, Wan Tang, Naiji Lu, Xin Tu, and Christopher M Homan. Connecting the invisible dots: reaching lesbian, gay, and bisexual adolescents and young adults at risk for suicide through online social networks. *Social science & medicine (1982)*, 69(3):469–474, August 2009.

[SDYH10] Noah Stupak, Nicholas DiFonzo, Andrew J Younge, and Christopher Homan. Socialsense: Graphical user interface design considerations for social network experiment software. *Computers in Human Behavior*, 26(3):365–370, 2010.

[SG03] Hope Jensen Schau and Mary C Gilly. We are what we post? self-presentation in personal web space. *Journal of consumer research*, 30(3):385–404, 2003.

[SKS12a] Adam Sadilek, Henry Kautz, and Vincent Silenzio. Modeling spread of disease from social interactions. In *Sixth AAAI International Conference on Weblogs and Social Media (ICWSM)*, 2012.

[SKS12b] Adam Sadilek, Henry Kautz, and Vincent Silenzio. Predicting disease transmission from geo-tagged micro-blog data. In *Twenty-Sixth AAAI Conference on Artificial Intelligence*, 2012.

[SOJN08] Rion Snow, Brendan O'Connor, Daniel Jurafsky, and Andrew Y Ng. Cheap and fast—but is it good?: evaluating non-expert annotations for natural language tasks. In *EMNLP '08: Proceedings of the Conference on Empirical Methods in Natural Language Processing*. Association for Computational Linguistics, October 2008.

[ST11] Cosma Rohilla Shalizi and Andrew C Thomas. Homophily and contagion are generically confounded in observational social network studies. *Sociological Methods & Research*, 40(2):211–239, 2011.

[SWC76] John Short, Ederyn Williams, and Bruce Christie. The social psychology of telecommunications. 1976.

[TBP+10] Mike Thelwall, Kevan Buckley, Georgios Paltoglou, Di Cai, and Arvid Kappas. Sentiment strength detection in short informal text. *Journal of the American Society for Information Science and Technology*, 61(12):2544–2558, 2010.

[TBP11] Mike Thelwall, Kevan Buckley, and Georgios Paltoglou. Sentiment in twitter events. *Journal of the American Society for Information Science and Technology*, 62(2):406–418, 2011.

[TG11] A. Tomas and K.J. Gile. The effect of differential recruitment, non-response and non-recruitment on estimators for respondent-driven sampling. *Electronic Journal of Statistics*, 5:899–934, 2011.

[TSSW10] A. Tumasjan, T.O. Sprenger, P.G. Sandner, and I.M. Welpe. Predicting elections with Twitter: What 140 characters reveal about political sentiment. In *Proceedings of the Fourth International AAAI Conference on Weblogs and Social Media*, pages 178–185, 2010.

[TZS+12] Jie Tang, Yuan Zhang, Jimeng Sun, Jinhai Rao, Wenjing Yu, Yiran Chen, and ACM Fong. Quantitative study of individual emotional states in social networks. *Affective Computing, IEEE Transactions on*, 3(2):132–144, 2012.

[WBS+09] Christo Wilson, Bryce Boe, Alessandra Sala, Krishna PN Puttaswamy, and Ben Y Zhao. User interactions in social networks and their implications. In *Proceedings of the 4th ACM European conference on Computer systems*, pages 205–218. Acm, 2009.

[WCTS12] Wenbo Wang, Lu Chen, Krishnaprasad Thirunarayan, and Amit P Sheth. Harnessing twitter 'big data' for automatic emotion identification. In *Privacy, Security, Risk and Trust (PASSAT), 2012 International Conference on and 2012 International Confernece on Social Computing (SocialCom)*, pages 587–592. IEEE, 2012.

[Wej09] C Wejnert. An empirical test of respondent-driven sampling: Point estimates, variance, degree measures, and out-of-equilibrium data. *Sociological Methodology*, 39(1):73–116, August 2009.

[WF94] Stanley Wasserman and Katherine Faust. *Social network analysis: Methods and applications*, volume 8. Cambridge university press, 1994.

[WLJH10] Jianshu Weng, Ee-Peng Lim, Jing Jiang, and Qi He. Twitterrank: finding topic-sensitive influential twitterers. In *Proceedings of the third ACM international conference on Web search and data mining*, pages 261–270. ACM, 2010.

[WS98] Duncan J Watts and Steven H Strogatz. Collective dynamics of small-worldnetworks. *nature*, 393(6684):440–442, 1998.

[ZHVGS10] Tauhid R Zaman, Ralf Herbrich, Jurgen Van Gael, and David Stern. Predicting information spreading in twitter. In *Workshop on Computational Social Science and the Wisdom of Crowds, NIPS*, volume 104, pages 17599–601. Citeseer, 2010.

[ZTS$^+$10] Yuan Zhang, Jie Tang, Jimeng Sun, Yiran Chen, and Jinghai Rao. Moodcast: Emotion prediction via dynamic continuous factor graph model. In *Data Mining (ICDM), 2010 IEEE 10th International Conference on*, pages 1193–1198. IEEE, 2010.

In: Social Networking
Editors: X. M. Tu, A. M. White and N. Lu
ISBN: 978-1-62808-529-7

Chapter 6

SOCIAL NETWORKING: ADDRESSING AN UNMET NEED IN THE YOUNG HAEMOPHILIA POPULATION

***Kate Khair*[1], *Mike Holland*[2*] *and Shawn Carrington*[2]**
[1]Great Ormond Street Hospital for Children NHS Foundation Trust, London, UK
[2]Haemnet, Battersea, London, UK

ABSTRACT

Haemophilia is an inherited bleeding disorder that is characterized by a deficiency of clotting factor, resulting in uncontrolled bleeding. Until the advent of modern management approaches, children born with haemophilia could expect frequent joint disease and hospitalisation, and a shortened life expectancy. Access to modern treatments allows most adolescents with haemophilia to manage their haemophilia at home, with near normal life expectancy and quality of life. But it has also reduced the potential for peer support and experiential learning from older peers. Those who rarely see the consequences of poor adherence with treatment may ultimately find it more difficult to follow health care advice and manage their disease. Affected adolescents already face relative geographical isolation due to the disease rarity, and many experience the "social isolation" that arises from having a disease that requires active management. Social networking, aided by modern communication technologies, may offer health benefits to adolescents with haemophilia and bleeding disorders by reducing isolation and enhancing access to peer support.

I. INTRODUCTION

Haemophilia is an X-linked hereditary bleeding disorder caused by a deficiency of clotting factor VIII (FVIII) (in haemophilia A) or factor IX (FIX) (in haemophilia B). The clotting factor deficiency arises as the result of mutations in the respective clotting factor genes. Haemophilia has an estimated frequency of approximately one in 10,000 births.

---

[*] www.haemnet.com

Estimates based on annual surveys conducted by the World Federation of Hemophilia indicate that the number of people with haemophilia worldwide is about 400,000. [1]

Haemophilia A is more common than haemophilia B, and accounts for 80-85 per cent of the haemophilia population. Haemophilia occurs in all racial groups globally. Despite the hereditary nature of the condition, around one third of cases occur spontaneously, with no previous family history. [2]

The severity of bleeding in haemophilia is generally correlated with the clotting factor level (Table 1). Individuals with severe haemophilia are at high risk of repeated joint and muscle bleeds that can lead to progressive joint damage and long-term disability. Without treatment, there is also significant risk of intracranial bleeding and its severe complications, including permanent cerebral damage and death. Treatment with appropriate factor replacement therapy can stop or prevent bleeding episodes reducing the associated complications, improving quality of life and normalising life expectancy.

Table 1. Relationship of bleeding severity with clotting factor level

Severity	Clotting factor level	Bleeding episodes
Severe	<1 IU/dL (<0.01 IU/mL) or <1% of normal	Spontaneous bleeding into joints or muscles, predominantly in the absence of identifiable haemostatic challenge
Moderate	1–5 IU/dL (0.01–0.05 IU/mL) or 1–5% of normal	Occasional spontaneous bleeding; prolonged bleeding with minor trauma or surgery
Mild	5–40 IU/dL (0.05–0.40 IU/mL) or 5 to <40% of normal	Severe bleeding with major trauma or surgery. Spontaneous bleeding is rare

Until the advent of modern management approaches, children with severe haemophilia spent considerable time in hospital and some even attended special schools. [3] In these situations, affected boys had the opportunity to build relationships with others affected by the same disease. Access to modern treatments now allows most adolescents to manage their condition at home. While there are clear benefits to transferring most treatment from the hospital setting to the home, this has also resulted in fewer opportunities for children to meet and interact with peers facing the same health challenges. As a rare disease, and one that is increasingly managed away from traditional health care settings, it is unsurprising that several adolescents with haemophilia have alluded to feelings of social and geographic isolation in conversations with members of their comprehensive care team and during focus group discussions.

Living with a long term condition can be a significant source of stress and anxiety, particularly for people with haemophilia who must acquire considerable knowledge and management skills at a young age. [4] Those who understand their disease and its treatment are typically less anxious and better able to manage their health both physically and emotionally. [5] In addition to the health expertise provided by formal sources, patients living with chronic illness often depend on friends, relatives and other patients for support and practical advice. This experiential knowledge may be lacking in adolescents with haemophilia who now have less opportunity to interact with similarly affected peers, and particularly so in those who have no affected family members to approach for advice.

One way to address the current dispersion of adolescents with haemophilia across geographies and cultures would be to provide opportunities to connect via online networks. Social networking and digital communication technologies are already central features of most adolescents' lives and may offer health benefits to those with rare chronic health conditions by reducing isolation, enhancing access to peer support and encouraging positive health choices. [6,7] Social networks have the potential to provide an infrastructure that is currently lacking to allow young people with haemophilia to overcome geographical and cultural barriers to information seeking and sharing and to restore the practical support networks that used to exist for this specific patient population.

II. Social Networking in Health

The Broader Landscape

In the UK alone, there are currently more than 50,000 organisations that supply health and social care information to the public. [8] A recent Pew Research Center study reported that 80 per cent of internet users search for health information online, although some have suggested these numbers may be lower for adolescents. [9] Peer-to-peer sites are far less prevalent than one-way health information sites, but healthcare social networking continues to grow and has been shown to be a powerful tool for promoting healthy choices and behaviours, by equipping motivated individuals with the ability to share information and learn from the experiences of others. [10,11,12]

Prior to analysing the available information sources and social networks available to adolescents with haemophilia, we explored the broader landscape of interactive health communities. These range from simple forums to more advanced infrastructures. Inspire (www.corp.inspire.com) and Patients Like Me (www.patientslikeme.com) have constructed sophisticated data capture platforms and work with pharmaceutical industry partners to identify patients for clinical trials and interpret data submitted by patients brought together via the sites. Most other two-way conversation sites provide an open platform for discussions and are funded either by advertising sponsors, venture capital investment or individual pharmaceutical companies. Across disease areas, communities established for the purpose of sharing health experiences tend to be most common in the areas of diabetes and cancer and include heavily visited sites such as Discuss Diabetes (www.discussdiabetes.com), The Diabetes Hands Foundation (www.diabeteshandsfoundation.org) forums, Diabetic Connect (www.diabeticconnect.com), and Breast Cancer Care (www.breastcancercare.org.uk/ community).

Although these successful communities are characterised by broad engagement and active conversations, adolescents with haemophilia differ from patients with type 2 diabetes and cancer in that most patients with these conditions were diagnosed as adults. Patients with type 1 diabetes and cystic fibrosis bear more similarities to the haemophilia population in that they have grown up with their diagnosis and adolescence represents a pivotal period during which the responsibility for care shifts from the parents to the individual.

There are currently a number of online cystic fibrosis communities (Table 2), each of which takes a slightly different approach to connecting and supporting patients. Although the prevalence of cystic fibrosis is approximately three times higher than haemophilia, [13] young people with cystic fibrosis often remain physically isolated from peers due to concerns over spreading communicable bacteria, particularly *Pseudomonas aeruginosa* and *Burkholderia cepacia.* [14,15] Despite this caution, the need for peer-to-peer support remains and has led parents and healthcare professionals to encourage social networking in the adolescent cystic fibrosis population in place of summer camps and other face-to-face meetings.

Table 2. Online cystic fibrosis communities

Site	Approach	URL
Cystic Life	Lets children connect and browse questions	www.cysticlife.org
CF Voice	Education and connection divided by age	www.cfvoice.com
Club CF	Rewards patients for submitting stories	www.clubcysticfibrosis.com
Rock CF	Range of programmes to support CF patients and raise awareness	www.letsrockcf.org

Social Networking and Patient Support in Haemophilia

As with the cystic fibrosis community, there is potential for social networking to provide a framework for peer-to-peer support and practical experience sharing in the young haemophilia community. In general, adolescents with long-term conditions are a particularly at-risk population for reduced treatment adherence as they become more independent, spend more time away from home and begin to associate with different peer groups. One haemophilia nurse described this time as “starting from ground zero” in terms of adherence and awareness of potential complications of poor management.

A recent paper published in *Science* stressed the importance of peer-to-peer social networking sites in influencing healthy behaviours. [16] The need for adolescent-centred electronic resources to address educational needs prior to transition for patients with haemophilia has been voiced by health care professionals, who have called for interactive websites with features identified by young people as being desirable, such as games, animations, messaging and chat features. [17]

We undertook a desktop audit of websites offering information and two-way communication for adolescents with haemophilia, with a main focus on those intended for the UK population. There are several sites these are principally aimed at young people and families, predominantly sponsored by the pharmaceutical industry. Among these sites, we found only one that focuses specifically on the adolescent haemophilia population in the UK.

Questival (www.questival.com) is a UK membership-based site created by Bayer Healthcare. It allows young people to ask age-relevant questions they may not feel comfortable asking in a formal medical setting. Suggested topics include alcohol, smoking and drugs, herbal supplements, body art and sex. Access to the site is restricted to those given a passcode by their treatment centre staff. This limits the potential user base to a relatively small group of boys with severe haemophilia. Bayer has a similar non-membership website aimed at an adolescent haemophilia audience in the United States. The Living Beyond Hemophilia (www.livingbeyondhemophilia.com) site functions as a practical resource for young people as they prepare for higher education and start careers, offering a range of general pragmatic advice and specific considerations for young people with haemophilia as they begin to live independently. While the site itself does not support functionality to directly connect users, it does stress the importance of mentoring and peer support within the haemophilia community and encourages users to contact their local National Hemophilia Foundation chapter or a number of recommended mentoring organisations. However, other than specific groups on Facebook, we did not identify any sites that offered two-way communication for adolescents with haemophilia in the UK.

Adolescents with haemophilia have access to a number of on and offline sources of information on their conditions, but a strong peer-to-peer community serves a broader purpose than just supplying information. The World Federation of Hemophilia guidelines [1] recommend extended members of the comprehensive care team should strive to:

- provide as much information as possible about the physical, psychological, emotional, and economic dimensions of haemophilia, in terms the patient/parents can understand
- be open and honest about all aspects of care
- allow patient/parents to work through their emotions and ask questions
- talk to affected children, not just their parents. Children can often understand a good deal about their illness and can work with the physician if properly informed and educated
- encourage patients to engage in productive and leisure activities at home and in the workplace

An online community, with reasonable moderation to screen out misinformation and inappropriate conversations, could function as an extension of the comprehensive care team to provide support in each of these recommended areas. A national or international network also has the potential to uncover regional differences in management and raise awareness about haemophilia management in countries where treatment and prophylaxis are difficult to obtain.

In addition to limited opportunities for peer interaction, home treatment has led to decreased exposure to the consequences of poor treatment adherence. When, in the past, young people frequently visited treatment centres, they would have been exposed to those who had suffered severe joint complications permanently affecting mobility as a result of inadequate treatment in the past. Allowing the population to reconnect via a social network could subtly raise awareness about the risks of poor management and demonstrate through example the importance of following health care advice and management plans.

III. Learnings from Focus Groups and Primary Research

In addition to surveying the broader health website landscape and current options for peer-to-peer connection in haemophilia, we have also conducted primary research to assess the need and opportunity for developing a specific network for this population. In a series of focus groups and through online and paper-based questionnaires, we have explored how adolescents with severe haemophilia currently use social networks and if they would find an online peer health network beneficial. [18]

Insights from these focus groups indicated ubiquitous but varied use of social networks, a shift towards accessing the internet on mobile devices, and a general lack of peer support in the young haemophilia population. Although all members of the focus groups used social networking to some extent, their levels of participation varied, with most boys suggesting they are consumers of content rather than generators. One boy noted, "I used to use Facebook, but I haven't logged on in about a year because it's sort of boring now. People posting the same things," while another observed "I used to go on [Facebook] just to keep up with friends and what they're doing and events. But I don't write anything myself."

In digital strategy, this behaviour of participation inequality is commonly described by the 90-9-1 principle and was first noted in the context of online social networks in the 1990s. [19] It describes the phenomenon that in most online communities, 90 per cent of users are observing but not posting content, 9 per cent are occasionally contributing or commenting on existing conversations and only 1 per cent of users are actively posting original content and starting discussions and therefore tend to account for the majority of contributions. Participation inequality is common to nearly all web communications and age groups and should be taken into account when considering activity in an online community as representative of the group's overall views or level of engagement with the site. Limited conversation should not necessarily be interpreted as limited engagement, and metrics for measuring reach should be carefully considered. For example, number of page views or time spent on the site would be more indicative of overall engagement than comments written in response to a blog post.

Another key aspect of social networks highlighted in the quotes above is the role they play in maintaining relationships with a group of friends. The one-to-many quality of social networking communications facilitates keeping groups of friends in contact as all members of the group can view the content posted by a few active members and maintain a social connection established at one point in time. One boy described how he maintained friendships that may have dissipated in the days before connecting online was so prevalent: "I've just been on a German exchange, and it was good because, I could just speak to my exchange free, and also we had an exchange group so everyone on the exchange from our school and the German school could all be in one group, and could speak and get involved in chats."

In 2011, the Cystic Fibrosis Trust conducted a study of online friendships in the UK, which found the average Brit has 55 physical friends and 121 online friends. [20] The CF Trust surveyed 3000 people age 16-86 and found that over a quarter feel as strongly connected or more connected to their online friends than to friends they socialise with in person. Commenting on the importance of online friendships specifically for people with health issues, Consultant Psychologist Helen Oxley noted, "For most people, the internet is a way of keeping in touch with loved ones and friends but for people who are isolated due to illness, it plays a more vital role and can often act as a lifeline." The study also found that people are far more likely to ask for an email address to connect via a social network than to ask for a phone number to keep in touch (23 per cent vs 5 per cent).

These findings correlate with comments from our focus groups, where participants noted they primarily connect on social networks with people they have already met. This apparent contradiction is consistent with the finding that people have more friends online than in person. Lower barriers of connection, also confirmed in the CF Trust study, mean that people are more likely to connect online with acquaintances than to maintain and develop those relationships in person. For young people with haemophilia, establishing online peer friendships that would otherwise be lost would be especially beneficial for times when they face challenges in coping with their condition and would like to reach out to someone else who understands. Indeed, some boys commented they do currently discuss bleeds privately online with other young people with haemophilia they have met through face-to-face events.

In addition to assembling acquaintances in a social network, we have explored the potential for making initial connections online. In one survey, 50 teenagers from around the country indicated that they would value the opportunity of being able to speak to other teenagers with haemophilia to get advice from them at times. [18] When asked about meeting others with haemophilia online, participants indicated they were not aware of any specific haemophilia online communities and would be unlikely to search for others with the condition in more general social networks. One boy noted, "I wouldn't really know how to search for a person with haemophilia. I don't really think they'd advertise, 'Follow me because I've got haemophilia.'"

When more specifically asked whether there had been points when they would have like to talk to someone facing the same health issues, one participant shared the following story: *"I've had that quite a few times. Recently, I had to stop playing basketball, because I was at quite a high level of competition ... it was affecting my health quite a lot because I was playing pretty much two or three hours a day, and that was quite stressful... I had to stop playing it because it was affecting my joints. And then I kind of thought would there be anyone out there who's had that kind of problem, and who's been through it, but I never thought of trying to like go on Facebook to find anyone or anything. But if ever I did, I guess I'd kind of bump into them."*

This reflection raises the important point that adolescents with haemophilia struggle with unique issues and would benefit from peer support from others who have faced the same challenges. Guidelines, healthcare professionals and parents can all direct what patients should do medically, but practical recommendations on coping with practical limitations are fundamentally different from advice from an authority figure. One boy noted he did not visit an informational site designed for adolescents, also mentioning that his mother had registered for it.

The UK's Haemophilia Society serves a vital purpose in the broader community and has a dynamic Facebook group with nearly 2000 members engaging in dozens of active conversations. The site and organisation are a valuable resource to individuals, friends and families affected by haemophilia, and especially to parents of affected children. Adolescents, however, have indicated that this strong parental and adult presence acts as a deterrent for them becoming directly involved and that they might be more likely to seek out peer support in a separate online space.

Sport and athletic activity are important to the young haemophilia community. In one focus group, nearly all the boys stated their main interest outside of school was some sort of organised team sport or athletic endeavour. The topic of sport is somewhat controversial with different treatment centres and health care professionals advising different levels of caution to patients, and young people with haemophilia must each find their own balance between keeping fit to protect joints and muscles and prudent choice of activity to avoid potential injury. A peer-to-peer network could be effective in both encouraging sports activity and also in helping each patient learn their own limits when it comes to sport.

In response to this research and the apparent unmet need for peer-to-peer support among boys with haemophilia, the authors of this paper have established SixVibe (www.sixvibe.com). This is an online membership-based community for young people with haemophilia and other bleeding disorders, consisting of a social networking platform, information and games (Figure 1). SixVibe is a professionally designed website, funded by unrestricted educational grants from multiple pharmaceutical industry partners, as well as donations from the charitable sector, and is endorsed by the UK Haemophilia Society. SixVibe is allied to a nurse-based professional network (www.haemnet.com) and the site moderators are able to generate information for patients at the request of haemophilia nurses, and to ensure the validity of those materials by "road testing" on nurses before they are released to adolescents.

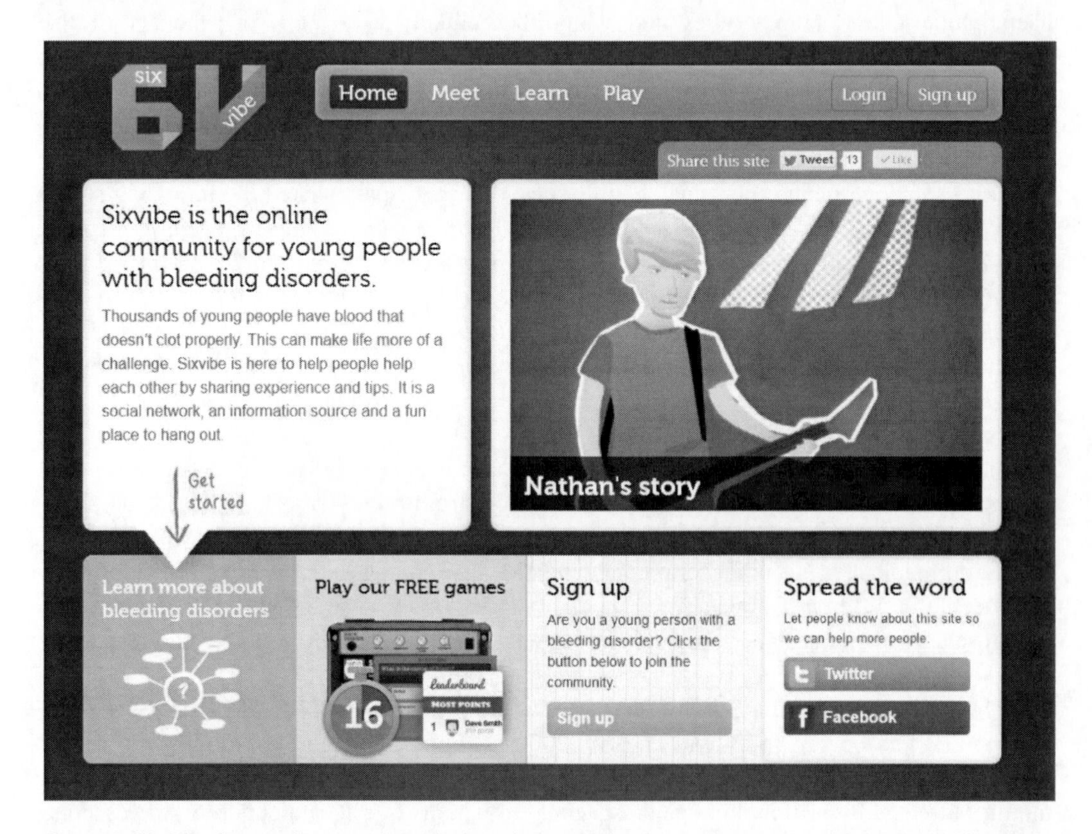

Figure 1. SixVibe is an online membership-based community for young people with haemophilia and other bleeding disorders.

In line with best practice guidance, SixVibe is fully moderated to help protect user safety and to keep the site free from parents and health care professionals. In the year since it was established site membership has grown substantially, with initial recruitment based predominantly on focus groups and recommendations from haemophilia nurses. This initial phase has identified a real need for a cross-community social networking platform. Future objectives will focus on encouraging active user engagement through strategic sharing of relevant conversation-starting content and motivating patients to share experiences and to support each other in making healthy choices.

Consideration

As with participation in online communities, our focus groups represent only a very small percentage of the UK haemophilia population. These patients tended to be from relatively urban settings and had universal access to the technology required for social networking. All the boys in focus groups had smartphones allowing internet access: in one focus group meeting, every boy indicated that their first activity each morning was to "reach for the phone."

Clearly, it is unlikely that all patients currently will have access to the same technology as the participants in our focus groups. We also recognise the growing trends towards ubiquitous connectivity, immediate communication and a shift from accessing online resources via desktop computers to accessing them via smartphones. Mobile devices and social media use continue to grow exponentially with mobile device sales reaching 440.5 million units in just the third quarter of 2011 and 42 per cent increase in smartphone sales from 2010. [21]

Consumer-facing businesses worldwide are redefining the meaning of customer engagement as the balance of power shifts from marketers communicating by one-way broadcast to empowered consumers expecting dialogue and response. Medical systems are also gradually transitioning to leverage the power of online resources, digital tools and empowered patients, especially in chronic conditions requiring lifelong management. This has exciting implications for improving the care of young people with haemophilia born into a hyperconnected world that only continues to bring more and more people together.

IV. Next Steps in Social Networking for Haemophilia

Our background research has shown a growing number of online communities providing support and connection for people with rare conditions. Adolescents with haemophilia in the UK have working knowledge of how to access general haemophilia information from both traditional and online sources but have suggested a gap in existing networks for peer support. Focus groups, surveys and anecdotal evidence have demonstrated that a customised social network for specifically designed for this age group and separate from existing initiatives involving parents and caregivers would provide the privacy necessary for engagement and would be beneficial to this population.

For any social network to achieve success, there must be sufficient content to hold user interest. Some focus group participants noted declining interest in Facebook because it had become "boring" and people had been "posting the same things." A social network for young people with haemophilia would have a smaller potential membership pool and would see much lower levels of user generated content than a site like Facebook. A smaller user base combined with the 90-9-1 rule of participation inequality might mean a specific community would struggle to meet the critical mass necessary to keep members engaged if it depended solely on user-generated content for material. Optimisation of the full range of available digital strategies may go some way towards overcoming this challenge.

A successful site for the young haemophilia population would likely contain a blend of user generated content and practical information about age-relevant topics, including travel considerations, sports, sex, living away from home and examples of how to explain haemophilia to new friends. Members would most likely be accessing the site for specific support and therefor engagement levels would be expected to be lower than a general social networking site.

In several other long-term conditions such as diabetes, asthma (www.myasthma.com) and HIV (www.myhiv.org.uk), initiatives are emerging that foster social networking with active community management and provide patients with practical apps to help them manage their conditions. In the haemophilia community, data collection in the form of detailed treatment diaries is a heavy burden that some patients view as a form of punishment. One boy in a focus group commented, "As a haemophiliac, it's not our fault that we have to do the injections in the first place ... and then we have to fill out a worksheet to say that we've done that, which seems like an extra chore, which adds to the idea like we're not normal."

Providing access to relevant digital management tools as an adjunct to a social network might help broaden the appeal of an online community by introducing it to users who might not initially be interested in discussing health issues with peers online. Digital tools could also improve adherence with data recording and treatment by leveraging the principles of gaming and "gamification". Several of the focus group participants noted they spend significant amounts of time playing digital games online. The principles of gaming could be leveraged to drive education about haemophilia, especially in the younger population.

The Wellcome Collection recently developed the flash-based game "Axon" to complement its exhibition "Brains: The Mind as Matter." The game drives education about the nervous system by challenging players to grow a neuron as long as possible by clicking on near-by nutrient sources. When a player runs out of time or paths to grow their axon, they receive a notification telling them what type of neuron they have grown based on the length, which links directly to the Wikipedia article describing that neuron. Axon was played over 1 million times in the first week and over 3.5 million times in the first month after launch and led to an exponential increase in page views of neuron-related articles on Wikipedia. [22]

Gaming to drive education differs slightly from the concept of gamification to change behaviour. Gamification is commonly defined as the process of using game design techniques in a non-game context, often to have a desired behavioural effect. It is becoming increasingly prevalent in health with thousands of apps and games currently on the market and under development. One particularly compelling example of gamification involves the use of video games with stroke patients. A team of game engineers, clinicians, and mathematicians based at Newcastle University developed "Circus Challenge" in collaboration with the Department of Health, the Wellcome Trust and the Health Innovation Fund. [23] The game encourages patients to perform physical activities, which can also be measured to determine if the patient is improving as a result of rehabilitation. [24]

Patients in our focus groups noted that employing the principles of gamification offered a potential strategy to encourage the completion of treatment diaries. One boy commented, "As a haemophiliac, you have to make a daily diary entry as to when you've done your injection. If you could incorporate that into a game, because I know myself it's really boring, it might make recording treatment more manageable." Haemophilia patients are currently required to record details of treatment administration, dosages, product codes, bleeds and activities. These data are required by healthcare commissioning authorities to justify treatment frequency and costs. Our focus group work indicated that few boys were particularly diligent in their data collection; several freely admitted to completing several months-worth of diary entries immediately prior to each scheduled consultation. Combining the principles of gamification with a digital tool to ease the data burden could have a substantial impact on the quality of treatment data collected by the adolescent haemophilia population.

In the longer term, social networks could also offer digital tools to capture and interpret patient reported outcomes, offer patients real-time personalised feedback on trackable disease management parameters and drive greater awareness of treatment disparities, both within health systems and globally. Developing pragmatic tools also capable of data collection offers an opportunity for industry partners to work collaboratively with patient populations both to give support and to gain insights that would be difficult to generate in small subsets within an already small population like the haemophilia community.

CONCLUSION

There is significant potential for social networking to bring together a currently isolated young haemophilia community. Adolescents are digital natives to social networking and currently lack a strong peer support network to discuss practical management issues of haemophilia. A successful online community would have to strike a careful balance between privacy, relevance and utility. In the longer term, digital tools could revolutionise treatment diary completion and overall chronic condition management.

ACKNOWLEDGMENTS

The authors wish to thank all of the boys who have completed questionnaires or attended our focus group meetings, the haemophilia nurses who have recruited participants or distributed questionnaires, and Marc Greenwood, Lara Oyesiku and Kristin Shine for their support and advice.

REFERENCES

[1] Srivastava A, Brewer AK, Mauser-Bunschoten EP, Key NS, Kitchen S, Llinas A, Ludlam CA, Mahlangu JN, Mulder K, Poon MC, Street A; Treatment Guidelines Working Group The World Federation Of Hemophilia. Guidelines for the management of hemophilia. Haemophilia. 2012 Jul 6. doi: 10.1111/j.1365-2516.2012.02909.x.

[2] http://www.who.int/genomics/public/geneticdiseases/en/index2.html#Haemophilia.

[3] Aronstam A, Rainsford SG, Painter MJ. Patterns of bleeding in adolescents with severe haemophilia A. Br Med J 1979; 1: 469-70.

[4] Civan A, McDonald DW, Unruh KT, Pratt W. Locating patient expertise in everyday life. GROUP ACM SIGCHI Int Conf Support Group Work 2009; 2009: 291–300.

[5] Barlow JH, Stapley J, Ellard DR, Gilchrist M. Information and self-management needs of people living with bleeding disorders: a survey. Haemophilia 2007; 13: 264–70.

[6] Stanton E, Lemer C. Networking for healthcare reform. J R Soc Med 2010; 103: 345–6.

[7] Valente TW. Network Interventions. Science 2012; 337; 49-53.

[8] http://www.nhs.uk/aboutNHSChoices/aboutnhschoices/Aboutus/Pages/the-information-standard.aspx.

[9] http://pewinternet.org/Reports/2011/HealthTopics/Summary-of-Findings.aspx

[10] Christakis NA. Social networks and collateral health effects. BMJ 2004; 329: 184–5.

[11] Farmer AD, Bruckner Holt CE, Cook MJ, Hearing SD. Social networking sites: a novel portal for communication. Postgrad Med J 2009; 85: 455–9.

[12] Brownstein CA, Brownstein JS, Williams DS 3rd, et al., The power of social networking in medicine. Nat Biotechnol 2009;27:888e90.

[13] http://www.who.int/genomics/public/geneticdiseases/en/index2.html#CF.

[14] Speert DP, Henry D, Vandamme P, Corey M, Mahenthiralingam E. Epidemiology of Burkholderia cepacia complex in patients with cystic fibrosis, Canada. Emerg Infect Dis 2002;8(2): 181-7.

[15] Gibson RL, Burns JL, Ramsey BW. Pathophysiology and management of pulmonary infections in cystic fibrosis. Am J Respir Crit Care Med 2003; 168: 918-51.

[16] Valente TW. Network Interventions. Science 2012; 337; 49-53.

[17] Breakey VR, Blanchette VS, Bolton-Maggs PHB. Towards comprehensive care in transition for young people with haemophilia. Haemophilia 2010; 16: 848–57.

[18] Khair K, Holland M, Carrington S. Social networking for adolescents with severe haemophilia. Haemophilia 2012; 18(3): e290-6.

[19] Hill WC, Hollan JD, Wroblewski D, McCandless T. Edit wear and read wear. Proceedings of CHI'92, the SIGCHI Conference on Human Factors in Computing Systems (Monterey, CA, May 3-7, 1992), pp. 3-9.

[20] www.cftrust.org.uk/VF_Press_Release_090411.doc

[21] “Gartner Says Sales of Mobile Devices Grew 5.6 Percent in Third Quarter of 2011; Smartphone Sales Increased 42 Percent.” Press Release, November 2011. http://www.gartner.com/it/page.jsp?id=1848514.

[22] http://preloaded.com/games/axon/.

[23] http://wellcometrust.wordpress.com/2012/05/24/limbs-alive-video-games-aid-stroke-recovery/.

[24] http://www.youtube.com/watch?feature=player_embedded&v=9HhzYZWbzp8.

In: Social Networking
Editors: X. M. Tu, A. M. White and N. Lu
ISBN: 978-1-62808-529-7

Chapter 7

MY BEST POTENTIAL FRIEND IN A SOCIAL NETWORK

Francisco Moreno*, Andrés González and Andrés Valencia
Escuela de Sistemas
Universidad Nacional de Colombia, Sede Medellín, Carrera

ABSTRACT

The online social networking phenomenon is growing rapidly all around the world. As a consequence, in recent years, several studies have been devoted to the analysis of social network (SN) sites. A specific issue that has been addressed is the identification of leaders based on well-known algorithms such as PageRank. The PageRank is an algorithm that has been applied to classify users of SNs and a leader is a user of the SN that obtain the highest PageRank. In this chapter, we propose a method to increase the PageRank of a SN user where we introduce a new concept: the *best potential friend*. Informally, the idea is to identify the user of a SN that would generate the highest increase in the PageRank of another user who is also member of the SN. We provide formal definitions, algorithms and some experiments for this subject.

Keywords: PageRank, best potential friend, social network, identification of leaders

1. INTRODUCTION

The online social networking phenomenon is growing rapidly all around the world. As a consequence, in recent years, several studies have been devoted to the analysis of social network (SN) sites. For example, the PageRank is an algorithm that has been applied to classify users of SNs, i.e., to compute user rankings [1]. The PageRank is a well-known algorithm to classify web pages [2], [3], [4]. Informally, the PageRank of a web page represents the probability that a web surfer, after a considerable period of time is visiting that

* Corresponding author: fjmoreno@unal.edu.co.

page. In addition to users classification in a SN, this algorithm has been extended [3] to identify, e.g., *competitivity groups*, i.e., sets of users that compete for having a better rank or *leadership groups*, i.e., the set of users that obtain the highest rank under certain conditions.

In this chapter, we propose a method to increase the PageRank of a SN user where we introduce a new concept: the *best potential friend*. Informally, the idea is to identify the user of a SN that would generate the highest increase in the PageRank of another user who is also member of the SN.

There are a few works that focus on how to increase the PageRank although they focus on web pages. Avrachenkov and Litvak [5], propose an optimal linking strategy. The idea is that a web page i should have only one outlink that points to a web page j such that j returns to i in just a few links. Although it is not practical that a web page has only one outlink, the idea is to link only to web pages that are relevant and belong to the same web community. The authors also show that, surprisingly, a new inlink is not always beneficial because the inlink may have irrelevant quality links. Kerchove, Ninove and Dooren [6] state that if a hyperlink from a page i to a page j is added, then the PageRank of j will increase. But in general, a webmaster does not have control on the inlinks of his webpage unless he/she pays (or makes an alliance) another webmaster to add a link to his/her page. However, it is natural to ask how a user could modify his PageRank by himself. This leads to analyze how the choice of the outlinks of a page can influence its own PageRank. Sydow [7] showed via numerical simulations that adding "well chosen" outlinks to a webpage may increase significantly its PageRank ranking. In [8] some advices are given to increase the PageRank. For example: to publish articles on other web pages relevant to a web page i can help to increase the PageRank of i if these articles include a link to i, to include a few outlinks to high quality sites (that creates a quality connection with web pages that search engines already consider to be quality sites [9]), to include XHTML in a web page because most search engines see a site as a text; whereas technologies such as Javascript, DHTML and Flash may make it hard for these search engines to crawl a site.

The chapter is organized as follows. In Section 2, we present the basic elements of the PageRank algorithm. In Section 3, we introduce the concept of best potential friend and present some experiments. Finally, in Section 4 we conclude the chapter and outline future work.

2. PageRank Algorithm: Basic Elements

Consider a SN with $n = 5$ users represented with a directed graph $GSN = (N, E)$, where N represents the set of nodes {1, 2, 3, 4, 5} and E the set of edges {(1,2), (1,3), (2,1), (2,4), (3,1), (4,2), (5,2)}, see Figure 1.

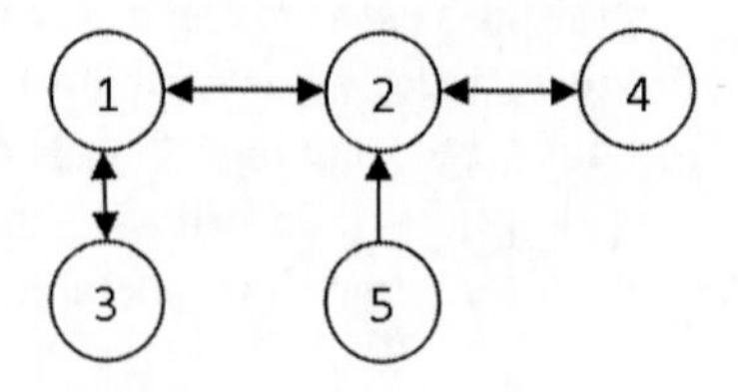

Figure 1. SN with 5 nodes, represented with a directed graph.

The goal is to classify the nodes (users) of the SN according to their links. To achieve this, we apply the PageRank method [1], [2].

To apply the PageRank method, first we build a *connectivity matrix* $H = (h_{ij}) \in R^{n \times n}$, $1 \leq i, j \leq n$, that represents the links of each node. If there exists a link from node i to node j, $i \neq j$, then $h_{ij} = 1$, otherwise $h_{ij} = 0$; if $i = j$ then $h_{ii} = 0$, see Figure 2.

From H matrix we build the *row stochastic matrix* $P = (p_{ij}) \in R^{n \times n}$, $1 \leq i, j \leq n$. A matrix is row stochastic if the sum of the elements of each of its rows is 1. P is calculated by dividing each element h_{ij} by the sum of the elements of row i of H, see Figure 3. Note that in a SN it is reasonable to assume that each user has at least one friend, i.e., there do not exist *dangling nodes* [11]; as a consequence, this sum may not be zero.

The PageRank method requires that the P matrix, in addition to be row stochastic, must be *primitive*. A non-negative square matrix is primitive [11] if the number of distinct eigenvalues of the matrix whose absolute value is equal to the spectral radius $\rho(P)$ is 1, where $\rho(P)$ is the maximum value in absolute value [10] of its eigenvalues. In order to ensure this property (and preserving the row stochastic property), we apply the following transformation [10].

$$G = \alpha P + (1 - \alpha) e v^{T}.$$

where G is known as *Google matrix* [10]. α is a *damping factor*, $0 < \alpha < 1$, and represents the probability with which the surfer of the network moves among the links of the H matrix, and $(1- \alpha)$ represents the probability of the surfer to randomly navigate to a link which is *not* among the links of H. Usually, α is set to 0.85, a value that was established by Brin and Page, the creators of the PageRank method [2], [10]. In [12], [13], and [14] the effect of several values of α is analized.

$$H = \begin{pmatrix} 0 & 1 & 1 & 0 & 0 \\ 1 & 0 & 0 & 1 & 0 \\ 1 & 0 & 0 & 0 & 0 \\ 0 & 1 & 0 & 0 & 0 \\ 0 & 1 & 0 & 0 & 0 \end{pmatrix}$$

Figure 2. Connectivity matrix.

$$P = \begin{pmatrix} 0 & 1/2 & 1/2 & 0 & 0 \\ 1/2 & 0 & 0 & 1/2 & 0 \\ 1 & 0 & 0 & 0 & 0 \\ 0 & 1 & 0 & 0 & 0 \\ 0 & 1 & 0 & 0 & 0 \end{pmatrix}$$

Figure 3. Row stochastic matrix.

On the other hand, $e \in R^{n\times 1}$ is the vector of all ones and $ev^T = 1$. v is called *personalization* or *teletransportation* vector and can be used to affect (to benefit or to harm) the ranking of the nodes of the network [10]: $v = (v_i) \in R^{n\times 1}$: $v_i > 0$, $1 \le i \le n$. Usually, $v = (1/n)$ and is known as *basic personalization vector*. However, if we want to affect (to benefit or to harm) the ranking of a specific node i, v may be defined as follows: Let $0 < \varepsilon < 1$ then $v_i = (v_{ij}) \in R^{n\times 1}$: $v_{ii} = 1 - \varepsilon$, $v_{ij} = \varepsilon/(n-1)$ for $i \neq j$. In this way, when ε is close to zero, the ranking of node i tends to increase, but if ε is close to one, its ranking tends to decrease. As a consequence, ε is usually set to 0.3; a value commonly used in the literature [1].

Note that the constraint $ev^T = 1$ allows us to define a v vector such that benefits (or harms) the ranking of several nodes simultaneously. For example, if v = (7/20 7/20 1/10 1/10 1/10) then the ranking of nodes 1 and 2 tend to be benefited whereas the ranking of nodes 3, 4, and 5 tend to be harmed. We denote PPR the Personalized PageRank as the PageRank of a node using some pre-scribed personalization vector and we denote PR_j the PageRank vector computed using a personalization vector v_j.

In Figure 4 we show the G matrix which was computed using $\alpha = 0.85$ and the basic personalization vector.

From G matrix we can compute the PageRank vector π as follows. We consider a system of equations $\pi^T = \pi^T G$, where π^T = [q1 q2 q3 q4 q5]. In addition, to ensure that π is a probability vector, we also consider the equation: q1 + q2 + q3 + q4 + q5 = 1.

For the running example, the system of equations is

$$0.0300\,q1 + 0.4550\,q2 + 0.8800\,q3 + 0.0300\,q4 + 0.0300\,q5 = q1 \quad (1)$$

$$0.4550\,q1 + 0.0300\,q2 + 0.0300\,q3 + 0.8800\,q4 + 0.8800\,q5 = q2 \quad (2)$$

$$0.4550\,q1 + 0.0300\,q2 + 0.0300\,q3 + 0.0300\,q4 + 0.0300\,q5 = q3 \quad (3)$$

$$0.0300\,q1 + 0.4550\,q2 + 0.0300\,q3 + 0.0300\,q4 + 0.0300\,q5 = q4 \quad (4)$$

$$0.0300\,q1 + 0.0300\,q2 + 0.0300\,q3 + 0.0300\,q4 + 0.0300\,q5 = q5 \quad (5)$$

$$q1 + q2 + q3 + q4 + q5 = 1 \quad (6)$$

We solved the system using MATLAB; results are showed in Table 1. Results show that node 2 has the highest PageRank whereas node 5 has the lowest.

$$G = \begin{pmatrix} 0.0300 & 0.4550 & 0.4550 & 0.0300 & 0.0300 \\ 0.4550 & 0.0300 & 0.0300 & 0.4550 & 0.0300 \\ 0.8800 & 0.0300 & 0.0300 & 0.0300 & 0.0300 \\ 0.0300 & 0.8800 & 0.0300 & 0.0300 & 0.0300 \\ 0.0300 & 0.8800 & 0.0300 & 0.0300 & 0.0300 \end{pmatrix}$$

Figure 4. G matrix computed using $\alpha = 0.85$ and the basic personalization vector.

Table 1. PageRank Vector π

Node	*PageRank*	
1	0.3073	
2	**0.3313**	→ Highest PageRank
3	0.1606	
4	0.1708	
5	**0.0300**	→ Lowest PageRank

As a second example, we compute the PageRank vector π with the personalization vector of node 3, i.e., PR_3 with $\varepsilon = 0.3$, i.e., $v_3 = (3/40\ 3/40\ 7/10\ 3/40\ 3/40)$. The corresponding G matrix is showed in Figure 5.

The system of equations is

$$9/800\ q1 + 349/800\ q2 + 689/800\ q3 + 9/800\ q4 + 9/800\ q5 = q1. \tag{7}$$

$$349/800\ q1 + 9/800\ q2 + 9/800 q3 + 689/800\ q4 + 689/800\ q5 = q2. \tag{8}$$

$$53/100\ q1 + 21/200\ q2 + 21/200\ q3 + 21/200\ q4 + 21/200\ q5 = q3. \tag{9}$$

$$9/800\ q1 + 349/800\ q2 + 9/800 q3 + 9/800\ q4 + 9/800\ q5 = q4. \tag{10}$$

$$9/800\ q1 + 9/800\ q2 + 9/800\ q3 + 9/800\ q4 + 9/800\ q5 = q5. \tag{11}$$

$$q1 + q2 + q3 + q4 + q5 = 1. \tag{12}$$

The resulting PR_3 vector is showed in Table 2.

$$G = \begin{pmatrix} 9/800 & 349/800 & 53/100 & 9/800 & 9/800 \\ 349/800 & 9/800 & 21/200 & 349/800 & 9/800 \\ 689/800 & 9/800 & 21/200 & 9/800 & 9/800 \\ 9/800 & 689/800 & 21/200 & 9/800 & 9/800 \\ 9/800 & 689/800 & 21/200 & 9/800 & 9/800 \end{pmatrix}$$

Figure 5. G matrix computed with $\alpha = 0.85$ and with the personalization vector v_3.

Table 2. PR_3 vector with $\varepsilon = 0.3$

Node	*PageRank*	
1	0.3391	→ Highest PageRank
2	0.2732	
3	**0.2491**	
4	0.1274	
5	0.0113	→ Lowest PageRank

Note that node 3 improved its ranking with respect to the PageRank vector of Table 1, because it changed from 0.1606 to 0.2491.

3. Best Potential Friend

3.1. Definition

We define the Best Potential Friend BPF of a node *i*, to be the node that when linked to *i*, provides the highest increase in the PageRank of *i*. That is:

Let $GSN = (N, E)$ be the initial graph. Let $\pi_i(GSN)$ denote the *i* component of the PPR for some personalization vector v . Given $i \in N$, let: $Z(i)=\{j \in N: i \neq j, (j,i) \notin E\}$, i.e., the set of nodes that *are not* linked to *i*. Let $E'(i,j)= E \cup \{(j,i)\}, j \in Z(i)$, be the initial set of edges *E* plus a new edge from *j* to *i*, and let $GSN'(i,j)=(N, E'(i, j))$. Then we say that $k \in Z(i)$ is the BPF of *i* if the following condition holds: $\pi_i(GSN'(i,k)) = \max(\pi_i(GSN'(i,j))), j \in Z(i)$.

That is, the BPF of a node *i* is the node *j* of the SN, $j \neq i$, such that if *j* becomes friend of *i*, *j* is the node that generates the highest increase in the PageRank of *i*. We define the Potential Friend PageRank Vector of a node *i* $PFPRV_i = \pi_i(GSN'(i,j)), 1 \leq j \leq n, j \neq i$.

Consider the SN of Figure 1 and Table 1, where the PageRank of node 5 is 0.03. In Table 3 we show the $PFPRV_5$. We can see the change in the PageRank of this node depending of the node that is connected to it. In this example, node 2 is the BPF of node 5.

The next algorithm computes the $PFPRV_i$. Let $PFPRV_i(k)$, $1 \leq k \leq n$, $k \neq i$, be the maximum value in $PFPRV_i$, then *k* is the BPF of *i*.

Algorithm PFPRV(i, H: n x n)
Input: i = The node to which we will find the PFPRV
H = Connectivity matrix
Output: PFPRV = Potential Friend PageRank Vector of node i

```
1.  j = 1;
2.  WHILE j <= n DO
3.     IF(H[j, i] = 0 AND j ≠ i) THEN
          //If there does not exist edge from node j to node i.
4.        auxH = H; //auxH is a copy of H matrix
5.        auxH[j, i] = 1 //Connect node j to node i.
6.        Compute PageRank vector using auxH matrix
7.        PFPRV[j] = PageRank[i] //Get PageRank of node i and store it in PFPRV
8.     END IF
9.  j = j + 1; //Next node
10. END WHILE
11. RETURN PFPRV
```

Table 3. PFPRV$_5$ in the SN with 5 nodes

Node to be connected to node 5	PageRank of node 5 after connecting the node
1	0,1055
2	**0,1313**
3	0,0859
4	0,0975

Using PFPRV algorithm we can create the Potential Friend PageRank Matrix (PFPRM), i.e., we compute PFPRV for each user of the SN, as we show in the following algorithm.

Algorithm PFPRM(H: n x n)
Input: H = Connectivity matrix
Output: PFPRM = Potential Friend PageRank Matrix

1. **FOR** i = 1 **TO** n
2. PFPRM [i] = PFPRV(i, H)
3. **END FOR**
4. **RETURN** PFPRM

3.2. Experiments

The PFPRM for the SNs of figures 1 (SN with five users) and 6 (SN with ten users) are showed in tables 4 and 5. Empty cells mean that the node to be connected to the node of interest is already its friend in the SN.

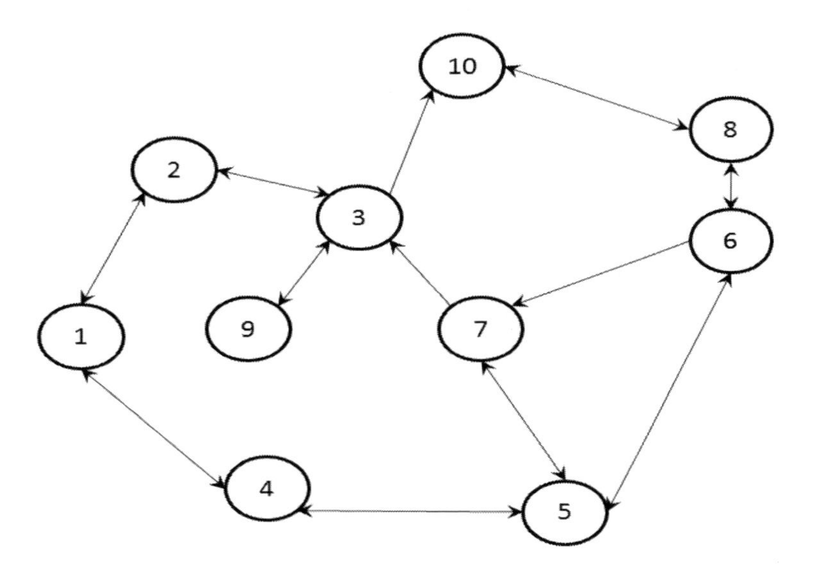

Figure 6. SN with 10 nodes, represented with a directed graph.

Table 4. PFPRM for the SN with 5 nodes

Node of interest	Node to be connected				
	1	**2**	**3**	**4**	**5**
1				0,3656	0,3193
2			0,3861		
3		0,2486		0,2332	0,1762
4	0,2524		0,2391		0,1813
5	0,1055	0,1313	0,0859	0,0975	

Table 5. PFPRM for the SN with 10 nodes

Node of interest	Node to be connected									
	1	**2**	**3**	**4**	**5**	**6**	**7**	**8**	**9**	**10**
1			0,1244		0,1185	0,1203	0,1184	0,1388	0,1169	0,1495
2				0,1133	0,1178	0,1186	0,1142	0,1375	0,1106	0,1494
3	0,1532			0,1555	0,1547	0,1554		0,1788		0,1958
4		0,1117	0,1208			0,1120	0,1120	0,1289	0,1130	0,1382
5	0,1395	0,1452	0,1512					0,1494	0,1433	0,1594
6	0,1343	0,1351	0,1355	0,1315			0,1324		0,1301	0,1294
7	0,1017	0,1046	0,1097	0,0977				0,1048	0,1016	0,1139
8	0,1761	0,1736	0,1682	0,1761	0,1752		0,1727		0,1646	
9	0,0783	0,0779		0,0781	0,0784	0,0783	0,0744	0,0962		0,1087
10	0,1417	0,1384		0,1426	0,1429	0,1401	0,1382		0,1300	

In order to facilitate the visualization and location of the BPF of each node in the SNs, results are showed in descending order, see tables 6 and 7. In bold, we show the BPF of each node (shaded row). For example, the BPF of node 1 in the SN with 5 nodes is node 4, see Table 6.

Note that results of Table 7 (SN with 10 nodes) show, e.g., that not necessarily the node with the highest PageRank in a SN is the BPF of a node. For example, the node with the highest PageRank in this SN is node 8; however, node 10 is the BPF of nodes 1, 2, 3, 4, 5, 7, and 9.

We also find the BPF for the ten worst ranking nodes in a SN of 769 nodes, Caltech-2005 [15], see Table 8. These nodes were 615, 717, 769, 671, 710, 494, 620, 583, 406, and 664. On the other hand, the nodes with the highest PageRanks in this SN are in descending order: 623, 207, 563, 60, 405, 88, 82, 411, 95, and 648.

Table 6. PFPRM for the SN with 5 nodes ordered by descending PageRank In parentheses node to be connected to the node of interest

Node of interest				
1	2	3	4	5
0.3656 (4)	**0.3861 (3)**	**0.2486 (2)**	**0.2524 (1)**	**0.1313 (2)**
0.3193 (5)		0.2332 (4)	0.2391 (3)	0.1055 (1)
		0.1762 (5)	0.1813 (5)	0.0975 (4)
				0.0859 (3)

Table 7. PFPRM for the SN with 10 nodes ordered by descending PageRank In parentheses node to be connected to the node of interest

Node of interest									
1	**2**	**3**	**4**	**5**	**6**	**7**	**8**	**9**	**10**
0.1495 (10)	**0.1494 (10)**	**0.1958 (10)**	**0.1382 (10)**	**0.1594 (10)**	**0.1355 (3)**	**0.1139 (10)**	**0.1761 (1)**	**0.1087 (10)**	**0.1429 (5)**
0.1388 (8)	0.1375 (8)	0.1788 (8)	0.1289 (8)	0.1512 (3)	0.1351 (2)	0.1097 (3)	0.1761 (4)	0.0962 (8)	0.1426 (4)
0.1244 (3)	0.1186 (6)	0.1555 (4)	0.1208 (3)	0.1494 (8)	0.1343 (1)	0.1048 (8)	0.1752 (5)	0.0784 (5)	0.1417 (1)
0.1203 (6)	0.1178 (5)	0.1554 (6)	0.1130 (9)	0.1452 (2)	0.1324 (7)	0.1046 (2)	0.1736 (2)	0.0783 (1)	0.1401 (6)
0.1185 (5)	0.1142 (7)	0.1547 (5)	0.1120 (6)	0.1433 (9)	0.1315 (4)	0.1017 (1)	0.1727 (7)	0.0783 (6)	0.1384 (2)
0.1184 (7)	0.1133 (4)	0.1532 (1)	0.1120 (7)	0.1395 (1)	0.1301 (9)	0.1016 (9)	0.1682 (3)	0.0781 (4)	0.1382 (7)
0.1169 (9)	0.1106 (9)		0.1117 (2)		0.1294 (10)	0.0977 (4)	0.1646 (9)	0.0779 (2)	0.1300 (9)
								0.0744 (7)	

Table 8. BPF for the ten worst ranking nodes in the SN of 769 nodes In parentheses node to be connected to the node of interest

Node of interest									
615	**717**	**769**	**671**	**710**	**494**	**620**	**583**	**406**	**664**
(201)	(201)	(201)	(201)	(201)	(201)	(201)	(201)	(201)	(201)
(88)	(9)	(88)	(88)	(88)	(88)	(88)	(88)	(88)	(88)
(9)	(60)	(9)	(60)	(9)	(9)	(9)	(9)	(9)	(9)
(60)	(207)	(60)	(207)	(60)	(207)	(60)	(60)	(60)	(60)
(207)	(72)	(207)	(72)	(207)	(72)	(72)	(207)	(207)	(207)
(72)	(82)	(72)	(82)	(72)	(82)	(82)	(72)	(72)	(72)
(82)	(563)	(82)	(563)	(82)	(563)	(563)	(82)	(82)	(82)
(563)	(95)	(563)	(95)	(563)	(95)	(95)	(563)	(563)	(95)
(95)	(648)	(95)	(648)	(95)	(648)	(648)	(95)	(95)	(648)
(648)	(51)	(648)	(51)	(648)	(51)	(51)	(648)	(648)	(51)
(51)	(623)	(51)	(623)	(51)	(623)	(623)	(51)	(51)	(623)
(623)	(176)	(623)	(176)	(623)	(60)	(176)	(176)	(118)	(118)

Note that although the node 623 is the node with the highest PageRank; it is not the BPF of any of the ten worst ranking nodes of this SN. However, node 623 appears in the top ten BPFs of eight of these nodes. Similarly, node 207 (the second node best PageRanked in this SN) appears in the top ten BPFs of nine of the ten worst ranking nodes. This suggests that although the nodes with the highest PageRanks are not necessarily the BPF of other nodes, they can be considered *good potential friends*.

CONCLUSION AND FUTURE WORK

In this chapter, we proposed an algorithm based on the PageRank method to find the BPF of a node. Results showed that the user of a SN with the highest PageRank is not necessarily the BPF of other nodes, i.e., when linked, it does not always provides the highest increase in the PageRank of other nodes.

As future work we consider the following. Suppose a user *u* cannot get the friendship of his BPF *w*; it would be interesting to analyze how it is affected the PageRank of *u* if it became friend of the friends of *w*. In other words, to analyze what happens to the PageRank of *u* if the friends of his BPF become his friends. We also plan to identify among the current links of a user *u*, i.e., its current friends, which is the most important one because when disconnected, it causes the highest decrease in the PageRank of *u*. Another work is to determine the number of new friends that a user needs in order to become the node with the highest PageRank in a SN.

REFERENCES

[1] Pedroche, F. (2010). Ranking nodes in Social Network Sites using biased PageRank, Instituto de Matemática Multidisciplinar, Universitat Politecnica de Valencia, Valencia, Spain.

[2] Page, L., Brin, S., Motwani, R. & Winograd, T. (1999). The PageRank Citation Ranking: Bringing Order to the Web. *Stanford Digital Library Technologies Project.* Available in: http://dbpubs.stanford.edu:8090/pub/1999-66.

[3] Pedroche, F. (2010). Competitivity Groups on Social Network Sites, *Mathematical and Computer Modelling*, *52*, 1052-1057.

[4] Newman, M. E. J. (2010). Networks: An introduction. Oxford University Press.

[5] Avrachenkov, K. & Litvak, N. (2004). *The Effect of New Links on Google PageRank*, technical report. INRIA.

[6] Kerchove, C., Ninove, L. & Dooren, P. (2011). *Maximizing PageRank via outlinks.* CESAME, Université catholique de Louvain.

[7] Sydow, M. (2005). Can one out-link change your PageRank?, Advances in Web Intelligence (AWIC 2005), *Lecture Notes in Computer Science*, vol. *3528*, Springer, Berlin, 408–414.

[8] Avrachenkov, K. & Litvak, N. (2004). *Decomposition of the Google PageRank and optimal linking strategy*, technical report, INRIA. Available in: http://www.inria.fr/rrrt/rr-5101.html.

[9] Bekman. S. *12 cosas que hacer para mejorar Rank de su sitio en Google*. Available in: http://stason.org/articles/money/seo/google/12_things_to_do_to_improve_your_site_google_page_rank.html#ixzz28lvmUDSh

[10] Pedroche, F. (2008). *Métodos de cálculo del vector PageRank.* Boletín de la Sociedad Española de Matemática Aplicada, *39*, (2007), 7-30.

[11] Gracia, J. M. (2002). Matrices no negativas, paseos aleatorios y cadenas de Markov, technical report, Universidad del País Vasco.

[12] Becchetti, L. & Castillo, C. (2006). The Distribution of PageRank Follows a Power-Law only for Particular Values of the Damping Factor WWW 2006, May 22-26, Edimburgo, Scotland.

[13] PageRank Explained with Javascript. Available in: http://williamcotton.com/pagerank-explained-with-javascript.

[14] Boldi, P. (2005). TotalRank: ranking without damping. *14th international conference on World Wide Web*. Chiba, Japan, 898-899.

[15] Traud, A., Kelsic, E., Mucha, P. & Porter, M. (2009). *Community structure in online collegiate social networks*, American Physical Society, APS March Meeting.

In: Social Networking
Editors: X. M. Tu, A. M. White and N. Lu
ISBN: 978-1-62808-529-7

Chapter 8

THE ROLE OF OPINION LEADERS AND INTERNET MARKETING THROUGH SOCIAL NETWORKING WEBSITES

Viju Raghupathi and Joshua Fogel*
Department of Finance and Business Management,
Brooklyn College, Brooklyn, NY, US

ABSTRACT

In the information age, Internet social media is integral for bridging the gap between companies and consumers. Companies use online communication from opinion leaders in the form of opinions, recommendations, suggestions and experiences on products, services, or sellers, in order to influence consumer decision making. This chapter reviews the empirical research from scholarly journals on what is known about online opinion leadership and the impact or potential impact on purchase or adoption decisions, intentions to purchase or adopt, or following product recommendations, using social networking websites (e.g., Facebook, LinkedIn). The search terms of, "(opinion leader OR opinion leaders OR opinion leader* OR influential people OR influential OR market mavens OR maven OR mavens OR key players OR key player*) AND (Internet OR online) AND (social network OR social network websites OR social network* OR social networking web* OR Facebook OR LinkedIn)" were searched in the numerous databases from the years 2001 through June 2012. Six relevant articles were reviewed from studies of opinion leadership. Opinion leadership in Internet social media marketing is analyzed in terms of online diffusion via electronic word-of-mouth (eWOM), identification of opinion leaders using various approaches, and factors that impact adoption and diffusion of opinion leadership. In addition to synthesizing and integrating the relevant literature, the chapter offers implications for future research and guidelines for marketers to effectively use social networking websites for consumer purchase and adoption decisions.

Keywords: opinion leaders, word-of-mouth, social networking, Internet

* Corresponding author: Dr. Joshua Fogel, Brooklyn College of the City University of New York, Department of Finance and Business Management, 218A, 2900 Bedford Avenue, Brooklyn, NY 11210, USA, Phone: (718) 951-3857, Fax: (718) 951-4385, E-mail: joshua.fogel@gmail.com.

INTRODUCTION

Social networks are very common in cultural, technical, personal and professional environments. These networks vary in the extent of functionality that they offer to users and in the type of cultures that emerge from their usage (Boyd & Ellison, 2008). The basic premise behind all online social networks is to offer an online platform for people with shared interests, views and activities to connect with each other (Boyd & Ellison, 2008). Within this framework these online social networks offer a variety of information and communication tools such as mobile applications, photo sharing, and video sharing. With the plethora of websites, a social networking website has certain characteristics that distinguish it from other websites. A social networking website can be defined as one that allows users: (a) to create their own profiles; (b) to designate their information created as private or public; (c) to generate a list of users to connect with; and (d) to be able to browse the list of connections that are made accessible by the users with whom they connected (Boyd & Ellison, 2008).

Popular Social Networking Websites

The first social networking website, SixDegrees.com, was introduced in 1997. It offered users the ability to create profiles, list friends, and search the list of friends (Boyd & Ellison, 2008). Since that time there have been numerous websites created that serve both social and professional purposes. Between the years of 1997 to 2010 there were 1.5 billion users of social networking websites (Kreutz, 2009). In September 2012, the top 10 social networking websites based on the Alexa global traffic rank and the United States traffic rank from Compete and Quantcast data is shown in Table 1 (eBizMBA, 2012).

As shown in Table 1, Facebook, Twitter, and LinkedIn are the three most visited social networking websites. Another ranking site lists Facebook as first, but then lists the order of MySpace, Friendster, Bebo, Zorpia, Netlog, Hi5, PerfSpot, Orkut, and Badoo (TopTenReviews, 2012). Infographic Labs reports that Facebook accounts for one out of every five webpage views worldwide. On Facebook, there are 250 million photos uploaded every day and 2.7 billion likes every day. About 57% of the users are female. An average Facebook user spends about 20 minutes per visit on the Facebook website (Infographic Labs, 2012).

Social Networking Websites and Businesses

Social networking is commonly used in both personal and professional settings. Social networking websites such as Facebook and LinkedIn offer businesses services such as blogging, professional networking, collaborative interactions, media and entertainment, and product and service reviews (Kumar & Sundaram, 2012). Organizations can use social networking websites for recruiting and selecting professionals, broadcasting and advertising product and service information, and tracking and maintaining relationships with customers (Dyrud, 2011). Social media has the potential to contribute about 900 billion to 1.3 trillion in annual value, by improving the overall productivity of organizations through enhanced communication and collaboration (McKinsley Global Institute, 2012).When businesses use the Internet or Internet social media for commerce, it is referred to as electronic commerce or e-commerce. Today, the growth of e-commerce is about five times that of traditional retail or commerce (Hughes & Buekes, 2012).

Social Networking Websites and Marketing

In the context of e-commerce, the strategy of focusing on social networks as effective channels for marketing is referred to as Internet marketing or e-marketing (Chailom, 2012). Marketers take advantage of the relationships that develop between users of social networking websites in order to increase sales (Even-Dar & Shapirab, 2011; Momtaz, Aghaie, & Alizadeh, 2011). The analysis of inter-relationships enables targeted marketing and identification of new customers who are not visible in traditional marketing channels (Momtaz et al., 2011). Some instances of e-marketing that occur in social networking websites include word-of-mouth marketing, diffusion innovation, buzz marketing, and viral marketing (Chailom, 2012; Hill, Provost, & Vollinsky, 2006). Businesses adopt social networking websites as a mechanism for three-way communication from firm to customer, customer to firm, and customer to customer. Through this three-way communication, businesses learn about customer preferences, understand the level of acceptance of their products or services, and allow existing customers to disseminate messages about products or services to other potential customers (Kumar & Sundaram, 2012). One of the current trends evident in utilizing social networking websites for marketing is the adoption of social coupons by networks such as Groupon and LivingSocial.

Social Networking Websites and Opinion Leadership

The abundance of product and service information on the Internet offers a wide range of alternatives for decision making. The theory of bounded rationality indicates that for the process of rational decision making people have limitations in terms of time and cognitive ability on the amount of information they can process and evaluate (Simon, 1982). In the face of such limitations, people often base their decisions on the opinions of others with whom they trust and feel close. Such people who are trusted and with whom others feel close and then whose opinions are sought after are potential opinion leaders. Opinion leaders are those whom can influence the opinions (Hellevik & Bjorklund, 1991) or the decisions of others (Chan & Misra, 1990). Opinion leadership theory follows a two-step flow framework in which information flows from mass media to opinion leaders and from opinion leaders to the opinion seekers (Burt, 1999; Katz, 1957; Katz & Lazersfeld, 1955; Lazersfeld, Berelson, & Gaudet, 1948). In e-marketing, marketers consider opinion leaders as integral for information and knowledge dissemination.

Communication from opinion leaders in the form of opinions, recommendations, suggestions and experiences about products and services are used to influence consumer decision making (Raghupathi et al., 2009). Such communication that occurs online is referred to as electronic word-of-mouth (eWOM) (Cheung, Luo, Sia, & Chen, 2009; Hennig-Thurau, Gwinner, Walsh, & Gremler, 2004; Hennig-Thurau & Walsh, 2003; Kozinets, de Valck,

Wojnicki, & Wilner, 2010) or word-of-mouth (de Valck, van Bruggem, & Wierenga, 2009). The majority of word-of-mouth referrals are made by a small group of active online users (Trusov, Bodapati, & Bucklin, 2010).

In opinion leadership and social networking websites, three general categories of research exist (Momtaz et al., 2011). One category of research looks at the identification of opinion leaders based on structural characteristics of the network, such as topology and centrality (position) in the network. Centrality in a network refers to the leader who is positioned at the center of interaction and is well-connected to others (Iyengar et al., 2011; Keller & Berry, 2003; van der Merwe & van Heerden, 2009; Zhang, Wang, & Xia, 2010). Centrality further affects a person's willingness to share information (Nov & Wittal, 2009; Wasko & Faraj, 2005). A second category of research looks at the relational characteristics that relate to the interactions of people in the network. In particular, the relationship between the leader and the members is a good indicator of word-of-mouth marketing (Li & Du, 2011). Tie strength is used in estimating strong and weak ties as a measure of the relationship (Granovetter, 1973). A third category of research looks at the personal characteristics of opinion leaders. These characteristics relate to cultural, technical or personal attributes that leaders exhibit (Shah & Scheufele, 2006), including innovativeness (Lyons & Henderson, 2005), education (Keller & Berry, 2003), political and personal trust (Shah & Scheufele, 2006), knowledge of stocks (Spann, Ernst, Skiera, & Soll, 2009), and motivation to share information (Ho & Dempsey, 2010).

Another earlier body of literature on opinion leadership and general social networking (i.e., but not specifically Internet social networking) classifies opinion leadership identification methods as self-reporting or self-designating, sociometric, and key informant methods (Rogers & Cartano, 1962). The self-reporting method involves asking people about how influential they think they are (Childers, 1986; Flynn, Goldsmith, & Eastman, 1994; King & Summers, 1970). The sociometric method involves the respondents indicating from whom they obtained advice on a topic (Engel, Blackwell, & Miniard, 1987; Rogers & Cartano, 1962). The key informant method includes asking informed individuals, rather than all members of a community, to identify whom they think will be opinion leaders (Engel et al., 1987; Rogers & Cartano, 1962).

Table 1. Top 10 Social Networking Websites in September 2012

Rank	Social Networking Website	Estimated Unique Monthly Visitors
1	Facebook	750,000,000
2	Twitter	250,000,000
3	LinkedIn	110,000,000
4	MySpace	70,500,000
5	Google+	65,000,000
6	DeviantArt	25,500,000
7	LiveJournal	20,500,000
8	Tagged	19,500,000
9	Orkut	17,500,000
10	CafeMom	12,500,000

Also, studies on opinion leadership explore the factors that impact it in either a professional/work setting (Sparrowe, Wayne, & Kraimer, 2001) or leisure/non-work related setting (Herring et al., 2005). Only a few studies explore opinion leadership in a combined professional and non-work related setting, such as when one asks a colleague at work for advice on a personal issue (Raghupathi, Arazy, Kumar, & Shapira, 2009).

Opinion leaders, by view of their social standing, are perceived to be early adopters of innovation (Bilgram, Brem, & Voigt, 2008; Rogers, 1995; Summers, 1970). They serve as lead users – users who face the need for a new product or service before the actual customers do. In this way, opinion leaders increase the rate of acceptance of new products and services (Morrison, Roberts, & von Hippel, 2000).

The success of social media marketing is based on the efficiency of contributions of the users. This is, in turn impacted by the willingness and motivation to share information (Raghupathi et al., 2009; Wasko & Faraj, 2005). There are two parties in the opinion leadership process: the information seeker (i.e., the recipient who seeks advice or opinion) and the information source (i.e., the opinion leader who provides the advice or opinion) (Flynn et al., 1994). In order to exert influence on others in advice giving, opinion leaders need to instill trust (Nahapiet & Ghoshal, 1998). Trust is therefore a key factor in opinion leadership studies. Trust not only impacts the willingness of a source to give advice but also of the recipient to accept advice in online settings (McKnight, Choudhury, & Kacmar, 2002; Smith, Menon, & Sivakumar, 2005). Competence and benevolence of opinion leaders are some dimensions of trust (Levin & Cross, 2004; Raghupathi et al., 2009). Competence or expertise is the ability of the source to offer accurate information (Bristor, 1990; Childers, 1986; Gibbons 2004). Benevolence is the perception of the intentions of the source in giving the advice. If the seeker perceives the intention of the source to be harmful, the seeker would avoid getting advice or opinions from such a source (Levin & Cross, 2004; McKnight et al., 2002).

In a broad sense, applications of opinion leadership in business and marketing generally considers the development of measurement scale for identifying opinion leaders and its application in different areas related to marketing such as healthcare industry, political science and public communications (Burt, 1999; Flynn et al., 1994; Howard, Rogers, Howard-Pitney, & Flora, 2000; Locock, Dopson, Chambers, & Gabbay, 2001). This chapter offers a comprehensive review of online opinion leadership and its impact or potential impact on purchase or adoption decisions, intentions to purchase or adopt, or following product recommendations, through using social networking websites (e.g., Facebook, MySpace, LinkedIn). This review is conducted on all the data-based articles published in peer-reviewed journals on the topics of opinion leadership and social networking from January 2001 through June 2012.

METHOD

Inclusion and Exclusion Criteria

The relevant inclusion and exclusion criteria were determined prior to searching through the databases. Inclusion criteria for the articles were that the articles were either qualitative or quantitative with empirical data on opinion leadership in social networking websites that had the objective of impacting decisions on either: 1) purchase or adoption, or 2) the intention to purchase or adopt. The exclusion criteria for the reviewed articles were: 1) retrieved from non-peer reviewed journals, 2) non-empirical articles, or 3) not written in English.

Search Strategy

On July 3, 2012, a search was done from multiple databases for all relevant articles from the year of 2001 through the year of 2012. The terms used for the search included the multiple phrases of (opinion leader OR opinion leaders OR opinion leader OR influential people OR influential OR market mavens OR maven OR mavens OR key players OR key player) AND (Internet OR online) AND (social network OR social network websites OR social network* OR social networking web* OR Facebook OR LinkedIn). The Ebsco database platform was used to search the databases of Business Source Complete, Academic Search Complete, CINAHL, CINAHL Plus with Full Text, Communication & Mass Media Complete, Computers & Applied Sciences Complete, EconLit with full text, MEDLINE, PsycARTICLES, PsycINFO, and SocINDEX with Full Text. Additionally, the retrieved articles were perused to see if other relevant articles were referenced within those articles.

RESULTS

The search retrieved a total of 99 articles. Some of the articles retrieved were repetitive. Many were deemed not pertinent to the review focus, as although they studied social network topics, this did not specifically include social networking websites. Only six articles were deemed relevant and are included in the review.

Table 2 summarizes some relevant information from the six reviewed articles. The articles were published from the years of 2007 to 2012. Only three of the articles provided information on the year of data collection. Three articles were on the impact of opinion leadership using the personal characteristics of leaders and/or users. Three other articles focused on the characteristics of the social network within which the opinion leaders functioned. The location of where the empirical studies were conducted was specified only in two of the six articles. A reason for lack of location specification is that social networking websites are not necessarily restricted to one particular location as the Internet is worldwide.

Below are brief summaries of all articles that were included and reviewed. They are categorized based on the characteristics used in studying the impact of opinion leadership on the decision to purchase or intention to purchase.

Personal Characteristics

Acar and Polonsky (2007) studied 427 undergraduate college students from the Northeastern United States in 2006. The only overall demographic data provided was sex from the initial enrolled sample of 451 where 50.6% were women; 24 were excluded from the analyses since they did not have a Facebook social networking website profile. In this study, opinion leadership was measured with a 4-item Likert-type scale with items used by Childers (1986). A sample item included, “I often notice that I serve as a model for others.” Opinion seekers were defined as those subject to interpersonal influence with regard to product purchase. This was measured with a 3-item Likert-type scale with items used by Bearden, Richard, Netemeyer, & Teel (1989). The study found that opinion leadership had a significant negative association with time spent online on social networking websites. Opinion leadership did not have a significant association with brand communication. Opinion seekers had a significant positive relationship with time spent online on social networking websites.

Li and Du (2011) retrieved 294 accessible URLs by searching with the keyword of “Apple iPhone” in the social networking website of MySpace.com. They identified 311 blogs and 259 bloggers. The date when data were obtained was not provided. Demographic data of age, sex, and race/ethnicity were not provided. Opinion leaders were measured using the BARR method. The BARR method includes blog content, author properties, reader properties, and relationship between blog author and blog reader. The study found that among the blog content, author properties of both expertise and blog preference had a significant positive association with opinion leaders. With regard to blog reader content, blog preference had a significant positive association with opinion leaders. However, reader expertise had a significant negative association with opinion leaders.

Samutachak and Li (2012) studied 156 individuals from China participating for one year in the social networking website named Hao Kan Bu. The date when data were obtained was not provided. Demographic data of age, sex, and race/ethnicity were not provided. In this study, opinion leaders were defined as those with either a high degree of centrality or prominence. Centrality refers to the number of friends with whom a person has direct contact. This was created by adding together information from diary content, story content, and message content. Prominence refers to how often people contact this particular person. This was created by adding together information from diary content, story content, and message content. The outcome variables were word-of-mouth classified into three different types of positive, negative, and neutral. This was measured with a 9-item Likert-type scale. The study found that in the analyses for positive word-of-mouth, prominence and the interaction of centrality and prominence had significant positive associations while centrality did not have any significant association. In the analyses for negative word-of-mouth, prominence and the interaction of centrality and prominence had significant positive associations while centrality did not have any significant association. In the analyses for neutral word-of-mouth, centrality, prominence and the interaction of centrality and prominence all had significant positive associations.

Network Characteristics

Smith et al., (2007) conducted a study to determine opinion leadership based on the areas of influence and the size of the personal network. In this study, they surveyed 11,791 participants from the CNET Networks brand social networking website. The date when data were obtained was not provided. The only demographic data provided was sex from the sample where 27% were women. Opinion leadership was measured using the relationship between personal social network size and topics of interest. Personal network size for a person was defined as the number of people one communicated with at least once a month in the categories of close/personal friends, casual friends, neighbors, adult family members, co-workers / employees / work supervisors / clients, and church/civic organization members. Personal social network size was categorized as less connected (10 or fewer connections), moderately connected (11-99 connections), and highly connected (100 or more connections). Topics of interest included 20 different categories ranging from cars to technology. The study found that there were overall significant differences between the social network size groups. In the post-hoc analyses, those with highly and moderately connected network sizes had significantly greater mean number of topics of interest as compared to those with a less connected network size.

Lu et al., (2011) studied 571,686 users of the online social network website named delicious.com. They studied 1,675,008 directed links of which 169,378 pairs were reciprocal links. The data were collected in 2008. Demographic data of age, sex, and race/ethnicity were not provided. In this study, opinion leaders were identified using the LeaderRank algorithm. The LeaderRank algorithm measures the relationships between leaders and fans. Users are ranked based upon bidirectional links between node connections. The LeaderRank algorithm was compared to the PageRank algorithm used by Google. The study found that the LeaderRank algorithm was a better approach for identifying opinion leaders. Also, the PageRank algorithm rated as higher rank those nodes with only a small number of opinion leaders.

Table 2. Characteristics of opinion leadership studies reviewed

Reference	Year Data Obtained	Location	Approach	Topic
Acar & Polansky (2007)	2006	US	Personal characteristics	Opinion leadership, opinion seeking, brand leadership.
Li & Du (2011)	Unknown	Unknown	Personal characteristics	Identification of opinion leaders based on blog content, author properties, reader properties, and relationship between blog author and blog reader.
Samutachak & Li (2012)	Unknown	China	Personal characteristics	Centrality and prominence.
Smith et al., (2007)	Unknown	Unknown	Network characteristics	Identification of opinion leadership using personal social network size and topics of interest.
Lu et al., (2011)	2008	Unknown	Network characteristics	Identification of opinion leaders using LeaderRank algorithm.
Li et al., (2012)	2010	Unknown	Network characteristics	Identification of opinion leaders based on SEAD (Social Endorser-based Advertising) score.

Li et al., (2012) studied 312 participants in the three social network groups of students, office workers and random members. The date when data were obtained was not provided. Demographic data of sex and age were provided. Sex characteristics for women were 43% of the student group, 51% of the office worker group, and 40% of the random worker group. Age was reported with low and high distributions ranging from 20-32 years for the student group, 23-45 years for the office worker group, and 18-43 years for the random member group. Opinion leadership was measured using the Social Endorser-based Advertising (SEAD) framework. The SEAD framework for opinion leadership is the sum of the influence score, preference score and the rating-feedback score which are used to derive an overall endorser-rating score. The authors created an application in the Facebook social networking website where they used the SEAD framework. The influence score was measured using the Friends link within Facebook to determine the frequency of interactions (centrality). There were four types of centrality indicators. First, degree centrality is the number of direct connections between nodes. The out-degree approach of the number of network members from which an individual can receive a response was used to determine degree centrality. Second, betweenness centrality is the interaction between two nonadjacent nodes. Third, closeness centrality measures the minimum length of an indirect path from one user to another. Fourth, tie strength combines closeness centrality and frequency of contact between users. The preference score was measured by matching the preference for a product from the user receiving the advertisement sent by the endorser. The rating-feedback score was measured using a Likert type scale from offensive to joyful to assess the advertising message sent by the endorser. The top 10 users were classified as endorsers. The study found that the SEAD approach was the best approach as compared to a number of other approaches for number of advertisements delivered and also number of individual users reached for all three groups of students, office workers and random members.

CONCLUSION

Personal Characteristics

Among the three reviewed articles on personal characteristics, one study (Acar & Polonsky, 2007) found that opinion leaders had a negative relationship with time online on social networking websites. However, another study (Samutachak & Li, 2012) found that prominence, which refers to how many people contact the opinion leader, had a significant impact on positive word-of-mouth through social networking websites. This suggests that even if opinion leaders do not spend as much time online as others, their influence in terms of diffusing positive word-of-mouth is integral for marketing on social networking websites. Also, opinion leadership has a significant impact on negative word-of-mouth (Samutachak & Li, 2012), indicating the importance of identifying as early as possible opinion leaders who may be negatively inclined towards purchasing of products and services through social networking websites. Marketing approaches should target these opinion leaders to attempt to modify or change their approach toward the product or service. The significant impact of opinion leaders on positive word-of-mouth on social networking websites is further shown where one study reports that readers have a preference for blogs in which opinion leaders are members (Li & Du, 2011). However, on social networking websites expertise is negatively associated with opinion leaders (Li & Du, 2011). This suggests that people with prior expertise do not perceive opinion leaders as important for their purchase decision making.

In terms of centrality, Samutachak and Li (2012) showed that centrality alone does not have a significant impact on word-of-mouth on social networking websites except on neutral word-of-mouth, which implies that opinion leaders who are in a central position only pass on information without giving it a positive or a negative connotation. This finding is distinct from other classic studies of non-Internet based word-of-mouth research such as Katz (1957) that promote centrality or the number of contacts as a determinant of the influential power of opinion leaders, and Levin and Cross (2004) who propose that the more central a member's position in a network, the more willing a member would be to contribute knowledge. This suggests that companies looking to market a product or service on social networking websites may have a more effective approach by identifying and targeting opinion leaders with prominence rather than with centrality. The reason is that opinion leaders with a central position may just pass on neutral information without giving it a positive or negative spin.

Network Characteristics

Among the three articles reviewed on network characteristics, one study found positive associations between opinion leadership and greater number of areas of influence (Smith et al., 2007), indicating that opinion leaders often have influence in multiple areas rather than just a single area. These results correspond with the classical two-step model proposed by Katz and Lazersfeld (1955) that describes how information first is consumed by opinion leaders who may be experts in a certain field and then becomes diffused to users in other areas, through a system of attitudes and beliefs. Also, the reviewed results of the positive impact of the area of influence or expertise on opinion leadership on social networking websites (Smith et al., 2007) are consistent with previous research on opinion leadership (Childers, 1986; Gibbons, 2004). However, a later opinion (Roch, 2005) may disagree as this person wrote that the influence of opinion leaders is mainly because of the informational advantages they possess rather than from their expertise.

Also, in the area of identification of opinion leaders on social networking websites, Lu et al., (2011) show how critical it is to have algorithms that accurately identify influential users by analyzing the bidirectional links between node connections, so as to effectively spread and form opinions. Similarly, Li et al., (2012) further emphasize the importance of deploying a multi-attribute technique in identifying opinion leaders including influence, preference, and rating feedback from the other users. This is consistent with previous literature on opinion leadership (Childers, 1986; Myers & Robertson, 1972; Roch, 2005).

Limitations and Future Research

This literature review only found a small number of articles in the context of opinion leadership on social networking websites with regard to purchase or intention to purchase products/services. There is the opportunity for future research to investigate more aspects of personal and network characteristics, especially with studies that combine research questions on both personal and network characteristics. Also, to our knowledge, there is not much research that compares the impact of opinion leadership across different product categories. Previous studies either focus on intention to purchase a specific product or a generic intention to purchase any product in a non-specific manner. It would be interesting to explore the differences and similarities that arise in opinion leadership due to functional or product attributes in different categories.

REFERENCES

Acar, A. S. & Polonsky, M. (2007). Online social networks and insights into marketing communications. *Journal of Internet Commerce*, *6*(4), 55-72.

Bearden, W., Richard, G., Netemeyer, J. & Teel, E. (1989). Measurement of consumer susceptibility to interpersonal influence. *Journal of Consumer Research*, *15*(4), 473-481.

Bilgram, V., Brem, A. & Voigt, K. (2008). User-centric innovations in new product development – systematic identification of lead users harnessing interactive and collaborative online-tools. *International Journal of Innovation Management*, *12*(3), 419-458.

Boyd, D. M. & Ellison, N. B. (2008). Social network sites: definition, history and scholarship. *Journal of Computer-Mediated Communication*, 13, 210-230.

Bristor, J. M. (1990). Enhanced explanations of word of mouth communications: The power of relationships. *Research in Consumer Behavior*, *4*, 51-83.

Burt, R. S. (1999). The social capital of opinion leaders. *Annals of the American Academy of Political and Social Science*, *566*, 1-22.

Chailom, P. (2012). Antecedents and consequences of e-marketing strategy: Evidence from e-commerce business in Thailand. *International Journal of Business Strategy*, *12*(2), 75-87.

Chan, K. K. & Misra, S. (1990). Characteristics of opinion leaders: A new dimension. *Journal of Advertising*, *19*, 53-60.

Cheung, M. Y., Luo, C., Sia, C. l., & Chen, H. (2009). Credibility of electronic word-of-mouth: Informational and normative determinants of online consumer recommendations. *International Journal of Electronic Commerce*, *13*(4), 9-38.

Childers, T. L. (1986). Assessment of the psychometric properties of an opinion leadership scale. *Journal of Marketing Research*, *23*(2), 184-188.

de Valck, K., van Bruggem, G. H. & Wierenga, B. (2009). Virtual communities: A marketing perspective. *Decision Support Systems*, *47*(3), 185-203.

Dyrud, M. (2011). Social networking and business communication pedagogy: Plugging into the Facebook generation. *Business Communication Quarterly*, *74*(4), 475-478.

eBizMBA (2012). Top 15 most popular social networking sites. Available at http://www.ebizmba.com/articles/social-networking-websites. Accessed September 17, 2012.

Engel, J. F., Blackwell, R. D. & Miniard, P. W. (1987). *Consumer Behavior* (5th edition) Chicago, IL: Dryden Press.

Even-Dar, E. & Shapirab, A. (2011). A note on maximizing the spread of influence in social networks. *Information Processing Letters*, *111*(4), 184-187.

Flynn, L. R., Goldsmith, R. E. & Eastman, J. K. (1994). The King and Summers opinion leadership scale: Revision and refinement. *Journal of Business Research*, *31*(1), 55-64.

Gibbons, D. E. (2004). Friendships and advice networks in the context of changing professional values. *Administrative Science Quarterly*, *49*(2), 238-262.

Granovetter, M. (1973). The strength of weak ties. *American Journal of Sociology*, *87*(6), 1360-1380.

Hellevik, O. & Bjorklund, T. (1991). Opinion leadership and political extremism. *International Journal of Public Opinion Research*, *3*, 157-181.

Hennig-Thurau, T. & Walsh, G. (2003). Electronic word-of-mouth: motives for and consequences of reading customer articulations on the internet. *International Journal of Electronic Commerce*, *8*(2), 51-74.

Hennig-Thurau, T., Gwinner, K. Walsh, G. & Gremler, D. (2004). Electronic word-of-mouth via consumer-opinion platforms: what motivates consumers to articulate themselves on the internet. *Journal of Interactive Marketing*, *18*(1), 38-52.

Herring, S. C., Kouper, I., Paolillo, J. C., Tyworth, M., Wright, E. & Ning, Y. (2005), Conversations in the Blogosphere: An analysis from the bottom up. *Proceedings of the 38th Annual Conference on Systems Sciences*, HICSS'05, Hawaii.

Hill, S., Provost, F. & Vollinsky, C. (2006). Network-based marketing: Identifying likely adopters via consumer networks. *Statistical Science*, *21*(2), 256-276.

Ho, J. Y. C. & Dempsey, M. (2010). Viral marketing: motivations to forward online content. *Journal of Business Research*, *63*(9-10), 1000-1006.

Howard, K. A., Rogers, T., Howard-Pitney, B. & Flora, J. A. (2000). Opinion leaders' support for tobacco control policies and participation in tobacco control activities. *American Journal of Public Health*, *90*(8), 1282-1287.

Hughes, S. & Beukes, C. (2012). Growth and implications of social e-Commerce and group buying daily deal sites: The Case of Groupon and Livingsocial. *International Business and Economics Research Journal*, *11*(8), 921-934.

Infographic Labs (2012). Facebook User Statistics 2012. Available at http://ansonalex.com/infographics/facebook-user-statistics-2012-infographic/. Accessed September 17, 2012.

Iyengar, R., Van den Bulte, C., Eichert, J., West, B. & Valente, T. W. (2011). How social networks and opinion leaders affect the adoption of new products. *Marketing Intelligence Review*, *3*(1), 16-25.

Katz, E. (1957). The two-step flow of communication: An up-to-date report on a hypothesis. *Public Opinion*, *21*(1), 61-78.

Katz, E. & Lazersfeld, P. F. (1955). *Personal influence: the part played by people in the flow of mass communication*. Glencoe, IL: Free Press.

Keller, E. B. & Berry, J. L. (2003). *The influentials: One American in ten tells the other nine how to vote, where to eat, and what to buy*, New York: Simon and Schuster.

King, C. W. & Summers, J. O. (1970). Overlap of opinion leadership across consumer product categories. *Journal of Marketing Research*, *7*(1), 43-50.

Kozinets, R. V., De Valck, K., Wojnicki, A. C. & Wilner, S. J. S. (2010). Networked narratives: Understanding word-of-mouth marketing in online communities. *Journal of Marketing*, *74*, 71-89.

Kreutz, C. (2009). The Next Billion - The Rise of Social Networking Sites in Developing Countries. Available at http://www.web2fordev.net/component/content/article/1-latest-news/69-social-networks. Accessed June 19, 2009.

Kumar, V. & Sundaram, B. (2012). An evolutionary road map to winning with social media. *Marketing Research*, 24(2), 4-7.

Lazersfeld, P.F., Berelson, B. & Gaudet, H. (1948). *The people's choice: how the voter makes up his mind in a presidential campaign*. New York, NY: Columbia University Press.

Levin, D. Z. & Cross, R. (2004). The strength of weak ties you can trust: The mediating role of trust in effective knowledge transfer. *Management Science*, *50*(11), 1477-1490.

Li, F. & Du, T. C. (2011). Who is talking? An ontology-based opinion leader identification framework for word-of-mouth marketing in online social blogs. *Decision Support Systems*, *51*(1), 190-197.

Li, Y., Lee, Y. & Lien, N. (2012). Online social advertising via influential endorsers. *International Journal of Electronic Commerce*, *16*(3), 119-154.

Locock, L., Dopson, S., Chambers, D. & Gabbay, J. (2001). Understanding the role of opinion leaders in improving clinical effectiveness. *Social Science & Medicine*, *53*(6), 745-757.

Lu, L., Zhang, Y., Yeung, C. H. & Zhou, T. (2011). Leaders in social networks, the delicious case. *PLOS One*, *6*(6), e21202.

Lyons, B. & Henderson, K. (2005). Opinion leadership in a computer-mediated environment. *Journal of Consumer Behavior*, *4*(5), 319-329.

McKinsley Global Institute (2012). The social economy: unlocking value and productivity through social technologies. Available at http://www.mckinsey.com/insights/mgi/research/technology_and_innovation/the_social_economy. Accessed September 21,2012.

McKnight, H. D., Choudhury, V. & Kacmar, C. (2002). Developing and validating trust measures for e-commerce: An integrative topology. *Information Systems Research*, *13*, 334-359.

Momtaz, N. J., Aghaie, A. & Alizadeh, S. (2011). Identifying opinion leaders for marketing by analyzing online social networks. *International Journal of Virtual Communities and Social Networking*, *3*(1), 43-59

Morrison, P. D., Roberts, J. H. & von Hippel, E. (2000). Determinants of user innovation and innovation sharing in a local market. *Management Science*, *46*(12), 1513-1527.

Myers, J. H. & Robertson, T. H. (1972). Dimensions of opinion leadership. *Journal of Marketing Research*, *9*, 41-46.

Nahapiet, J. & Ghoshal, S. (1998). Social capital, intellectual capital, and the organizational advantage. *Academy of Management Review*, *23*(2), 242-266.

Nov, O. & Wittal, S. (2009). Social computing privacy concerns: antecedents and effects. *Proceedings of the 27th International Conference on Human Factors in Computing Systems*, 333-336.

Raghupathi, V., Arazy, O., Kumar, N. & Shapira, B. (2009). Opinion leadership: Non-work-related advice in a work setting. *Journal of Electronic Commerce Research*, *10*(4), 220-234.

Roch, C. (2005). The dual roots of opinion leadership. *Journal of Politics*, *67*(1), 110-131.

Rogers, E. M. (1995). *Diffusion of innovation*. (5th edition), New York: Free Press.

Rogers, E. M. & Cartano, D. G. (1962). Methods of measuring opinion leadership. *Public Opinion*, *26*(3), 435-441.

Samutachak, B. & Li, D. (2012). The effects of centrality and prominence of nodes in the online social network on word of mouth behaviors. *Journal of Academy of Business and Economics*, *12*(2), 125-148.

Shah, D. & Scheufele, D. A. (2006). Explicating opinion leadership: Non political dispositions, information consumption and civic participation. *Political Communication*, *23*, 1-22.

Simon, H. A. (1982). *Models of bounded rationality*. Volume 2, Cambridge, MA: MIT Press.

Smith, D., Menon, S. & Sivakumar, K. (Summer 2005). Online, peer and editorial recommendations, trust and choice in virtual markets. *Journal of Interactive Marketing*, *19*(3), 15-37.

Smith, T., Coyle, J. R., Lightfoot, E. & Scott, A. (2007). Reconsidering models of influence: The relationship between consumer social networks and word-of-mouth effectiveness. *Journal of Advertising Research*, *47*, 387-397.

Spann, M., Ernst, H., Skiera, B. & Soll, J. H. (2009). Identification of lead users for consumer products via virtual stock markets. *Journal of Product Innovation Management*, *26*(3), 322-335.

Sparrowe, R.T., Wayne, S. J. & Kraimer, M. L. (2001). Social networks and the performance of individuals and groups. *Academy of Management Journal*, *44*(2), 316-325.

Summers, J. O. (1970). The identity of women's clothing fashion opinion leaders. *Journal of Marketing Research*, *7*, 178-185.

TopTenReviews (2012). 2012 social networking websites comparison. Available at http://social-networking-websites-review.toptenreviews.com/. Accessed September 17, 2012.

Trusov, M., Bodapati, A. V. & Bucklin, R. E. (2010). Determining influential users in Internet social networks. *Journal of Marketing Research*, *47*, 643-658.

van der Merwe, R. & van Heerden, G. (2009). Finding and utilizing opinion leaders: Social networks and the power of relationships. *South African Journal of Business Management*, *40*(3), 65-76.

Wasko, M. M. & Faraj, S. (2005). Why should I share? Examining social capital and knowledge contribution in electronic networks of practice. *MIS Quarterly*, *29*(1), 35-57.

Zhang, Y., Wang, Z. & Xia, C. (2010). Identifying key users for targeted marketing by mining online social network. *Proceedings of the 24th Conference on Advanced Information Networking and Applications Workshops*, 644-649.

In: Social Networking
Editors: X. M. Tu, A. M. White and N. Lu
ISBN: 978-1-62808-529-7

Chapter 9

SOCIAL NETWORKS AND THE JOB SEARCH: A FOCUS ON PEOPLE WHO ARE ASKED TO PROVIDE JOB ASSISTANCE

***Lindsey B. Trimble*[1]*, Julie A. Kmec*[2] *and Steve McDonald*[3]**
[1]Stanford University, US
[2]Washington State University, US
[3]North Carolina State University, US

ABSTRACT

Considerable research has examined the role social network contacts play in matching job seekers with jobs. Most research on this subject tends to focus on the people who use social networks during their job search, leaving us with little understanding of the people who are asked to provide job assistance, who we call "contacts." Contacts are an important focus of empirical investigation because becoming employed is contingent on them sharing useful job-related resources. We shed light on contacts' role in the job search process by identifying the people who are likely to be asked to help with job searches and the conditions in which they help job seekers to receive offers. Drawing on data from a unique survey of contacts in Washington state, we find that the young, employed, and minorities are more likely to be asked to help with a job search than older, unemployed, or white contacts, and that employed, male contacts who are familiar with job seekers' work qualifications help more than unemployed, female or unfamiliar contacts. Analyses also show that white, familiar contacts who have not received favors from job seekers in the past are more likely than minority, unfamiliar contacts who have received favors to provide the kinds of help that result in job offers. Contacts that help job seekers secure offers do not offer different types of help than those whose assistance did not lead to offers. We conclude with a discussion of the implications of studying contacts for the social network and job search literatures.

The use of *social networks* refers to job seekers' informal reliance on people—friends, family members, acquaintances, employers, or coworkers—to obtain job leads and other job-related resources which helps them in their search for employment (Trimble and Kmec 2011). Employers also rely on the social networks of their current and past employees to spread the word about job opportunities, informal information about jobs, and simplify the complicated process of selecting one candidate from many. Social network use during the job attainment process is common—close to half of all jobs are found through contacts (Chapple 2006; Granovetter 1995; Holzer 1996; Marsden and Gorman 2001), and job seekers'—as well as employers'—reliance on social networks during the employment process is becoming more prevalent as the structure of work becomes increasingly insecure (Williams, Muller, and Kilanski 2012).

Much of the focus of existing literature has been on those using social networks (i.e., job applicants who use them when searching for employment or employers who use them to search for applicants). Less attention has been paid to the people who are asked to provide job assistance, who we refer to as "contacts" (for notable exceptions see Marin 2012; Smith 2005, 2007, 2010). Contacts are key players in the social network exchange though and deserve to be the focus of empirical investigation. When contacts are able and willing to share the "right" information with job seekers, the social network exchange can potentially connect job applicants to employers. By contrast, when contacts cannot provide the "right" type of information for job seekers or refuse to help them, social network exchanges are not likely to connect applicants to jobs. Thus, studying contacts offers unique insights into the success of job applicants, because they are the "invisible" hand that link job seekers with jobs (McDonald and Day 2010).

In this chapter, we make the case that scholars stand to gain from studying social network contacts, in addition to the job seekers who use them to find employment and the employers who rely on network referrals to recruit new employees. To date, most research examining the social network process pays no attention to contacts, in large part because data on contacts is hard to come by. The few existing studies of contacts tend to be limited to analyses of the attributes of contacts (e.g., their sex, race, ethnicity, etc.) whose help leads directly to job seekers' application for jobs or employment (e.g., Beggs and Hurlbert 1997; Fernandez and Mateo 2006; Fernandez and Sosa 2004; Kmec and Trimble 2009; Lin, Ensel, and Vaughn 1981; Smith 2000).

While contacts' attributes are important to consider, knowledge of them alone does little to shed light on the mechanisms driving the networking process. For example, we know very little about the frequency with which people are asked to help with job searches, whether some people are more likely to be asked for help than others, and what motivates contacts to help applicants (but see Smith 2005, 2007, 2010; Marin 2012). Moreover, in focusing on contacts whose assistance leads to employment, we know little about contacts whose help does not result in employment. By directly studying contacts—regardless of whether or not the help they provide results in employment for the job seeker they are assisting—scholars will better understand how social network contact use works for some, but not others (Ioannides and Loury 2004; Marsden and Gorman 2001; Trimble and Kmec 2011). In other words, studying contacts can tell scholars about job seekers' differential "returns on investments" to networking. Scholars also stand to make stronger policy suggestions on issues surrounding job searches and employment.

In response to the current gap in our understanding of contacts' role in the job search process, this chapter brings to bear new evidence on social network use in the job search process. Specifically, we present findings from a unique survey of contacts who were asked whether they provided assistance with a recent job search, the Social Networks and Employment Assistance Survey (SNEAS). We use this survey to accomplish two goals. Our first goal is to shed much needed light on the integral part contacts play in job seekers' searches by providing empirical evidence on the frequency with which people are asked to help with others' job searches, who agrees to help with job searches, whose assistance leads to job offers, and how contacts' roles differ in networking exchanges that lead to job offers from exchanges that do not. Our second goal is methodological: to illustrate the types of information that researchers can glean from directly studying contacts and to show how studies of contacts can advance scholars' understanding of the mechanisms that make social networking work.

The remainder of the chapter is organized as follows. First, we identify and discuss three benefits to using network contacts in the job search process. Following this, we discuss common methodological approaches taken in social network studies and describe how studies of social network contacts can help scholars identify how contacts link job seekers with employment opportunities. We then describe the SNEAS data set and present our analyses on contacts. We conclude the chapter with ideas for future research.

SEEKING AND RECEIVING HELP FROM CONTACTS

Job seekers who use social network contacts to search for work tend to be advantaged relative to job seekers who use other, more formal job search methods (i.e., direct application, newspapers and web advertisements, etc.). First, social network contacts can increase the number of job offers a job seeker receives since contacts tend to produce more job opportunities to which a job seeker can apply than formal methods (applying to a greater number of jobs is likely to lead to a greater number of job offers) (Granovetter 1995; Lin 1999; Marsden and Gorman 2001; Marsden and Hurlbert 1988). Obtaining job information from social network contacts also increases the range of opportunities available to a job seeker, especially when the job seeker's contacts come from a variety of settings and social backgrounds and provide job seekers with non-redundant job information (Burt 1992; Granovetter 1995; Yakubovich 2005).

Second, the information job seekers obtain from informal social network connections is different—and often better—than information that flows formally. For example, social network connections can reveal information on when a company will experience a labor demand, and hence, when to best time an application (Powell and Smith-Doerr 1994). Informal social network contacts can circulate information—e.g., how to dress, whose names to mention, and what questions to ask during the interview (see Fernandez and Weinberg 1997)—that can give networked job applicants an "edge" over applicants using formal procedures. Network ties often convey to job applicants information about job settings that is not communicated through formal means (e.g., the "culture" of the job setting) but that can help job applicants self-select out of the running for a job that would not be a good "fit" (Fernandez, Castilla, and Moore 2000).

A third way job seekers benefit from using social network contacts in the job search process is through the influence their contacts may have. For example, a contact might be in a position to hire applicants or advocate for the job seeker by putting in a good word for him or her (Trimble and Kmec 2011). This is especially true when the contact is employed at the place a job seeker is applying because the contact may be able to directly exert influence on the hiring of the applicant or indirectly by simply signaling an applicant's acceptability "by association." If the contact is well respected or occupies a high status in the company where the network contact applies, the contact's association with the job seeker might signal the applicant will be a good hire since we tend to assume that individuals are connected to people like them (Podolny 2005) and because a high status contact might have a lot at stake (e.g., a reputation, social status) by endorsing an applicant who is inadequate for the job (see Lin 1999; Marsden and Gorman 2001; Miller and Rosenbaum 1997; Mouw 2003).

Not all job seekers experience these benefits when they turn to their social network contacts for help with their job search—some may be unable to help, some may refuse, and others may provide poor "quality" help that does not result in a job offer. To be successful, job seekers must ask contacts who are able and willing to share "good" information, "go out on a limb" for them, and provide the kinds of help that leads to job offers. Scholars know very little about the factors that affect whether contacts do these things, in part, because previous data collection and analytic techniques simplify or overlook their role in the job matching process. In the following section, we review three common methodological approaches to studying social network use—"start with hire" studies (Fernandez and Weinberg 1997), organizational case studies, and studies of job leads—and describe how each approach affects scholars' understanding of the mechanisms linking social network contact use to work outcomes (e.g., employment).

METHODOLOGICAL APPROACHES TO STUDYING SOCIAL NETWORK CONTACT USE

The "Start with Hire" Approach

Most studies of the social network process examine job seekers who become employed with the help of social network contacts. Since these studies, by design, only examine successful job searchers, Fernandez and Weinberg (1997) refer to them as "start with hire studies." These studies tend to consider the characteristics of the contacts who help job seekers find work (e.g., their gender, race, employment status) and job-related outcomes (e.g., job attainment, pay, or later turnover) (see Elliot 2001; Kmec 2007; Kmec and Trimble 2009; Mencken and Winfield 2000; Lin1999, 2000; Smith 2000). Other "start with hire studies" examine how knowing particular types of contacts affects job outcomes, whether or not these contacts actually helped job seekers find work (see Green, Tigges, and Browne 1995; Johnson, Bienenstock, and Farrell 1999).

Due to data limitations, the "start with hire" approach has been the primary way scholars have studied the effects of social networks during the job search.[1] Despite its ubiquity, the "start with hire" approach is problematic for two reasons (Fernandez and Weinberg 1997). First, it oversimplifies the relationship between social network contact use and employment outcomes because it overlooks the fact that becoming employed with the help of a social network contact is a multi-step *process* (Fernandez and Weinberg 1997). That is, in order for a job seeker to find work, she asks a contact for help, the contact must be able and willing to

[1] To avoid a "start with hire" approach requires data on all jobs a job seeker applied for, the method whereby the job seeker was linked to a job, and the outcome for each job application.

provide the help she needs in order to advance her search, and the job seeker must use the contact's resources to apply for and be offered a job (see Figure 1).[2]

By studying those who are already employed, researchers gloss over these steps, leaving us with little understanding of the factors that affect each stage in this process.

Second, this method systematically excludes people who search for work through social network contacts, but who do not find a job (Fernandez and Weinberg 1997). As a result, we have little understanding of the difficulties job seekers face when using this search strategy and the factors that make this job search strategy work for some but not others. These omissions have consequential implications for scholars interested in identifying the underlying mechanisms that make social network contact use a beneficial search strategy for some job seekers. Examining how social network contact use affects hiring and other employment-related outcomes using data collected *after* workers become employed is also methodologically flawed because this research samples on the dependent variable (Fernandez and Weinberg 1997). Moreover, those who study the effects of social networks on post-hire outcomes overlook the influence that social networks can have on both pre- *and* post-hire outcomes (Fernandez and Weinberg 1997). Put another way, the labor market positions of workers who find jobs through social network contacts may have *already* been influenced by contacts in systematic ways. To sidestep these problems, Fernandez and Weinberg (1997) suggest that scholars compare job seekers who do and do not become employed using social network contacts. One way to do this is to study the applicant pool of work organizations.

Organizational Case Studies

Following Fernandez and Weinberg's (1997) lead, some researchers studied the recruitment and staffing practices of single work organizations (Fernandez, Castilla, and Moore 2000; Fernandez and Fernandez-Mateo 2006; Fernandez and Mors 2008; Fernandez and Sosa 2004; Fernandez and Weinberg 1997; Peterson, Saporta, and Seidel 2000). These case studies depart from past research by tracing the employment outcomes of entire applicant pools as they progress through organizations' screening and hiring pipelines. They tend to focus on job referrals—that is, when current employees recommend potential workers for jobs at their work organizations—and compare four groups of applicants: 1) referred applicants who were offered jobs, 2) referred applicants who were not offered jobs, 3) non-referred applicants who were offered jobs, and 4) non-referred applicants who were not offered jobs.

Organizational case studies have advanced the social network contact and job search literature by shedding light on how social network contacts can affect job seekers' likelihood of becoming employed (e.g., by sharing information about when and how to submit job applications) (Fernandez and Weinberg 1997). In addition, this research approach draws attention to the role organizations play in tempering the effects of social network contact use on workers' outcomes (Fernandez and Weinberg 1997). Because organizations vary in the degree to which they recruit through and value referrals from current employees (Neckerman

[2] Some job seekers receive help from their social network contacts without asking for it. "Non-searchers" are the targets of unsolicited job help and the focus of a growing body of research (McDonald 2010; McDonald and Elder 2006). Our study focuses exclusively on active searchers.

and Fernandez 2003; Marsden 1996), the effect of social network contact use on job applicants' outcomes depends on organizational contexts.

Although the organizational case study approach greatly improves upon the "start with hire" approach, it still paints an incomplete understanding of the mechanisms that make social network contact use work. For example, case studies are limited in their generalizability. Second, organizational case studies tend to examine contacts' provision of referrals (see Fernandez and Fernandez-Mateo 2006; Fernandez and Sosa 2004; Fernandez and Weinberg 1997). Yet, providing a referral is but one of many important ways social network contacts help job seekers with their searches; contacts can also provide information about job opportunities outside their own work organization, share information about jobs or employers that does not circulate through formal channels (i.e, the work culture of a particular organization), put in a good word for job seekers, use their workplace influence on behalf of job seekers to ensure that they are offered a job, or hire the job seeker directly (Bian 1997; Fernandez and Weinberg 1997; Marsden and Gorman 2001; Lin 2001; Powell and Smith-Doerr 1994; Royster 2003, Yakubovich 2005). To understand *how* social networks operate during the job search, scholars should examine the diverse ways contacts help job seekers.

Finally, organizational case studies follow the applicant pool at a particular organization and thus begin when job seekers apply for jobs. As a result, they offer no insights into the processes that lead up to job seekers' application and the influence that social network contacts have on these processes. By definition, both start with hire studies (Fernandez and Weinberg 1997) and organizational case studies focus exclusively on networking situations in which job seekers receive help from contacts and so they do not shed light on the factors that affect whether contacts help (see Figure 1). To better understand why social network contact use works for some but not others, scholars need to examine the role of social network contacts during the pre-application process.

Studies of Job Leads

A third but less common methodological approach to studying job seekers examines the quantity and quality of leads job seekers receive (for example see Huffman and Torres 2001; Lin and Ao 2008; McDonald 2011; McDonald, Lin, and Ao 2009). This research has tended to focus on group differences in the receipt of job leads (McDonald et al., 2009), how membership in particular kinds of networks affects the likelihood that people learn about opportunities (McDonald 2011), or the quality of leads they receive (Huffman and Torres 2001).

An advantage to studying job leads is that they capture networking that does and does not result in job seekers' application for jobs (Trimble and Kmec 2011). Thus, these studies offer a more complete picture of the frequency with which networking occurs and the types of people who are involved in network exchanges. Yet, because job lead studies tend to focus on people who receive job leads, we know little about the factors that affect whether contacts pass along information about job openings, the origins of those job leads, or how they decide which job opportunities they will share with which job seekers. Similar to organizational case studies, job lead studies are limited in that they focus on one way contacts help, by sharing information about job opportunities.

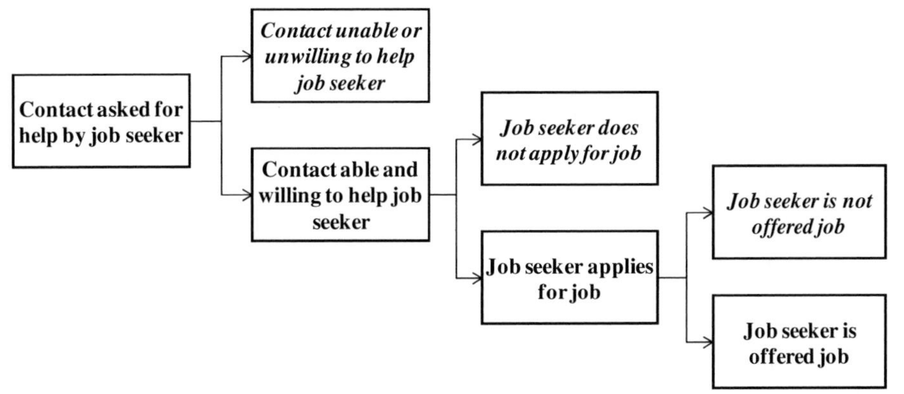

Italicized font shows when job seekers are unsuccessful at finding work through the social network contact.

Figure 1. Finding work through Social Network Contacts.

The Newly Emerging Focus on Social Network Contacts

As we noted in the introduction to this chapter, scholarship on contacts, in particular who they are, how they help, and the causes of variation in their helpfulness, is scant. Only recently have scholars turned their attention directly to social network contacts. This small body of research has revealed important attributes of contacts. First, several studies have found that not all contacts help when asked. In fact, some intentionally refuse to help job seekers (Marin 2012; Smith 2005, 2007, 2010). These studies have brought to light how the interpersonal relationships contacts share with job seekers impact their choice to help with others' job searches (Marin 2012; Newman 1999; Royster 2003; Smith 2005, 2007, 2010). For example, Smith (2010) found that black contacts placed tremendous value on trusting the job seekers who asked for their help—that is, they were unwilling to share their social resources with untrustworthy job seekers whom they feared would jeopardize their own reputations, but willing to help job seekers who exhibited pro-work behaviors (i.e., a strong work ethic). Although her research lacks generalizability to the broader population of job seekers, it points to the importance of identifying the stages that job seekers and contacts navigate early in the networking process. Moreover, her research demonstrates the value of studying contacts, rather than job seekers, for understanding the problems job seekers face when looking for work using their network contacts.

More recently, Marin (2012) studied the situations in which contacts shared unsolicited job information with the people in their network. She found that contacts were reticent to share information about job openings, and did so only when they were sure that the information would be wanted and appreciated. As a result, contacts tended to share information with "strong ties," people with whom they felt close, rather than with "weak ties." Both Smith (2005, 2007, 2010) and Marin's (2012) research demonstrates the importance of studying contacts directly. Contacts are not always forthcoming when asked for help with a job search, so understanding the factors that affect when and why they help is critical for identifying the mechanisms that make social network contact use in the job search work for some, but not others.

Despite the advances these recent studies have made, scholars still know very little about the frequency with which people are asked to assist with others' job searches and the kinds of people who are targeted to help with job searches. Nor do scholars know much about the contacts who help job seekers receive offers, or the ways they help them. Knowing this information is important for understanding job seekers' success with social networking because some contacts may be better at helping than others (Lin 2001; Newman 1999; Royster 2003, Smith 2005; 2007; 2010).

In an effort to shed light on this information, we draw on the SNEAS data to answer the following five questions:

1. How frequently are people asked to help with others' searches?
2. Who is asked to help?
3. Among those who are asked, who helps?
4. Among those who are asked, whose help leads to a job offer for the job seeker?
5. Is the type of contact assistance provided associated with the chance of receiving a job offer?

We begin by examining the frequency with which individuals are asked to help with someone else's job search. Then, to answer our second research question, we compare four characteristics—the sex, age, race, and employment status—of people who had and had not been recently asked to help with a job search. To address the third research question, we examine whether contacts that help with job searches have different characteristics than contacts that do not help. Because Smith (2005, 2007, 2010) and Marin's (2012) work shows that the strength of the relationship matters for contacts' provision of help, we also examine whether contacts who help have stronger relationships with job seekers than contacts who do not help. For our fourth research question, we explore whether the attributes of contacts who help job seekers receive offers differ from contacts who do not help contacts receive offers. Finally, we compare the types of job-related help that results in job offers with those that do not to answer our fifth research question.

DATA, VARIABLES, AND METHOD

Data

To examine these questions, we draw on data from the Social Networks and Employment Assistant Survey (SNEAS). During 2010-2011, the first author mailed an 11-page, 62-item paper questionnaire to a stratified, random sample of 1,850 Washington state households. The questionnaire was titled *"How do people look for work? An effort to understand how Washington residents help people find jobs."* The questionnaire asked respondents to share their views on a variety of topics including the state of the economy, perceptions about the possibility of finding work, and whether respondents had recently been asked to help with someone else's job search.[3]

[3] To determine whether respondents had recently been asked for help, the questionnaire included a nine-item question that asked whether a job seeker had recently sought any of the following from the respondent: 1)

Respondents who were recently asked to help with a job search, who we call "contacts," elaborated on the *last* time someone asked them for help. We refer to the last person who asked a contact for help as the "job seeker." Contacts described the types of help job seekers asked for, specifically, whether job seekers asked contacts to tell them about job openings, be a reference or provide a formal recommendation, introduce them to someone who could help with their search, or share information about a specific job or employer (e.g., the pay of the job or the culture at a particular work organization). Contacts explained whether they provided the help job seekers sought and then described the characteristics of the job seeker and their relationship with him or her. Finally, all respondents, whether or not they had recently been asked to help with a job search, were asked to answer a series of questions about their demographic characteristics. The questionnaire also asked contacts for their demographic information around the time when the job seekers asked for help.

To draw a sample, the first author contracted a private vendor with access to the United States Postal Service (USPS) Delivery Sequence File, a list of virtually all residential and commercial addresses to which the USPS delivers mail (Dillman, Smyth, and Christian 2009). The sample was chosen at random from lists of addresses stratified by urbanicity, and so half of the addresses are from urban blocks based on the 2000 United States Census while the other half are from rural blocks.[4] To ensure a random sample of individuals within households were selected to complete the questionnaire, the adult (age 18 years or older) in the household with the most recent birthday was instructed to complete the questionnaire (Battaglia et al., 2008).

The first author used the Tailored Design Method (TDM) to administer the survey (Dillman et al., 2009). This approach involved the use of multiple mailings, providing a $2.00 cash incentive for participation and a business reply envelope so that respondents could easily return the questionnaire, and tailoring mailings to increase interest among respondents (Dillman et al., 2009).[5] This methodology elicited a 34.4 percent response rate.[6]

To determine whether the SNEAS sample approximates the Washington state population, we compared the sex, age, race/ethnicity, and employment status of the SNEAS sample with

information about available job opportunities, 2) for respondents to talk with an employer on his or her behalf, 3) information about a specific job or employer, 4) advice about how to look for or apply for a job, 5) for respondents to directly hire him or her, 6) for respondents to be a reference or provide a formal recommendation, 7) advice on how to dress or how to act around an employer, 8) and introduction to someone who could help with their job search, or 9) for respondents to pick up or drop off a job application for him or her. We consider respondents who answered "yes" to any of these questions as having been recently asked to help with a job search.

[4] In other work, the first author compares the factors that affect whether urban and rural contacts help with job searches.

[5] Each household was contacted with four separate mailings. The first mailing consisted of a pre-notice letter which informed households that they had been randomly selected to take part in a survey about how people look for work and described the purpose and importance of the study. The second mailing, the questionnaire packet, was mailed approximately one week after the pre-notice letter. It included a cover letter that restated the purpose and importance of the survey in detail, the 11-page questionnaire, a $2.00 cash incentive, and a business reply envelope. One week after mailing the questionnaire packet, a postcard was mailed to all households which thanked participants who had already completed the questionnaire and reminded non-responders to do so. Finally, nearly one month after mailing the first questionnaire packet, non-responders were sent a replacement questionnaire packet. The replacement questionnaire packet was similar in content to the first packet; it included a cover letter which reiterated the significance of the study and reminded respondents to complete the questionnaire. Due to budget restraints, 500 households received two mailings (the questionnaire packet and postcard) instead of four and a $1.00 cash incentive instead of $2.00.

[6] We calculated this response rate using AAPOR's Standard Definitions response rate formula 6 (AAPOR 2010).

estimates of Washington state population using data from the 2009 American Community Survey (ACS 2009). This analysis revealed that the SNEAS sample has significantly more women, older people, and whites than the state of Washington as a whole. To ensure our analysis is representative of the population of Washington state, we created a weight that adjusts for sex, age, and urbanicity using the "raking" procedure in Stata. The raking procedure adjusts the sample one variable at a time to match the populations' proportions, and is the preferred method for constructing weights when population totals are unavailable for the variables one wants to weight by (Battaglia et al., 2004).[7]

Variables

Recently asked for help by a job seeker

We examine three outcome variables. The first measures whether respondents were recently asked to help with a job search—that is, whether they are "contacts." We coded SNEAS respondents as contacts if they reported that a job seeker had asked for at least one of nine types of job-related help in the three years preceding the survey (*contact*=1).[8] Respondents who had not been asked for any of these types of help are coded as "non-contacts" (*contact*=0).

Helped a Job Seeker

To determine whether contacts helped job seekers, we draw on a series of questions that measure whether job seekers provided help when asked for it. Recall that contacts were asked whether job seekers sought four "types" of job-related help: 1) information about job opportunities/openings, 2) information about a job or employer, 3) an introduction to someone who could help with their job search, and 4) a reference or letter of recommendation. We coded contacts who provided at least one of the types of help that job seekers needed as helping (*helped*=1). For example, consider the situation in which a contact is asked by a job seeker whether she knows of any job opportunities and also to be a reference. If the contact provides at least one type of help the job seeker needs—either sharing information about a job opportunity or providing the reference—we consider her to have "helped." Contacts who did not help job seekers—either because they did not know how to or because they decided not to—are coded as not helping (*helped*=0).

---

[7] We did not adjust the sample for race/ethnicity because the SNEAS sample has few racial/ethnic minorities. Disaggregating the sample by sex, age, urbanicity, and race/ethnicity resulted in many empty cells.

[8] Recall that these nine types of help include: 1) telling the job seeker about available job opportunities, 2) talking with an employer on the job seeker's behalf, 3) sharing information about a specific job or employer with the job seeker, 4) giving advice about how to look for or apply for a job, 5) hiring the job seeker, 6) being a reference or provide a formal recommendation, 7) giving advice on how to dress or how to act around an employer, 8) and introduction to someone who could help with their job search, or 9) for respondents to pick up or drop off a job application for him or her.

Helped a Job Seeker Receive an Offer

Finally, we measure whether contacts provided help that resulted in a job seeker's receipt of a job offer. The SNEAS questionnaire asked contacts whether the job seekers they helped applied for and were eventually offered jobs. Contacts who helped job seekers get at least one job offer are coded one for this variable (*offer*=1). Contacts who helped job seekers, but whose job seekers were not offered any of the jobs they were trying to get with contacts' help are coded zero (*offer*=0). Contacts who did not help job seekers (*n*=72), whose help did not result in job seekers applying for jobs (*n*=35), and who did not know whether the help that they provided job seekers eventually resulted in them applying for or being offered jobs (*n*=105) have missing values for this variable.

Respondents' Characteristics

We measure several characteristics of respondents to determine the types of people who are likely to be asked to help with others' job searches. *Respondent's sex* is measured using a dichotomous variable (1=female and 0=male). *Respondent's age* is measured with a series of dichotomous variables. Variables include people between the ages of 18-34 years, 35-44 years, 45-54 years, 55-64 years, and 65 years and older. *Respondent's race* is a dichotomous variable where 1=racial/ethnic minority and 0=white.[9] Finally, *employment status* is measured using a dichotomous variable that measures respondents' attachment to the labor force at the time of the survey (1=employed, 0=unemployed, retired, or an unemployed student).

Contacts' Characteristics

To examine differences between contacts that do and do not help with job searches, we measure the contact's characteristics at the time they were asked to help job seekers. *Contact's age* is measured with a series of dichotomous variables. Variables include people between the ages of 18-34 years, 35-44 years, 45-54 years, 55-64 years, and 65 years and older. *Contact's education* is an ordinal measure of his or her highest level of education at the time when the job seeker asked for help. Responses range from 1-5 where 1=less than a high school diploma or high school graduate (including GED), 2=some college, 3=2-year college degree, 4=4-year college degree, and 5=graduate or professional degree. *Contact's employment status* is a dichotomous variable measuring whether contacts were employed when job seekers asked for help and is coded 1 when contacts were employed and 0 when they were unemployed, retired, or an unemployed student.

[9] We coded respondents as racial/ethnic minorities if they described themselves as American Indian/Alaska Native, Asian, black or African American, Latino/Hispanic, or other (non-white), or if they selected two or more race categories (including white in conjunction with any other category). Eight respondents are American Indian/Alaskan Native, 26 are Asian, 6 are black or African American, 14 are Latino/Hispanic, no respondent reported being an "other" race, and 15 are two or more races (values are unweighted). Because so few respondents identified as racial/ethnic minorities, we aggregated the specific racial/ethnic groups into one variable that measures minority status.

Tie Strength

We also account for a number of variables that tap into the strength of the relationship between contacts and job seekers. First, we measure the length of a contact's relationship with a job seeker with a question that gauges how long contacts knew job seekers when they asked for help. *Years known* is an ordinal level variable and ranges from 1-4 with 1=less than one year, 2=one to two years, 3=more than 2 years, but less than 5 years, and 4=5 or more years. We also measure how close contacts felt to job seekers when they asked for help. Responses are coded: 3=very close, 2=somewhat close, 1=slightly close or not at all close. We include a measure of the contact's familiarity with the job seeker's work qualifications. Responses are coded: 3=very familiar to 1=slightly familiar to not at all familiar. Finally, to tap into whether contacts and job seekers had exchanged favors prior to job seekers asking contacts for help, we include a measure of whether job seekers had done contacts any favors, job-related or otherwise, in the past. *Helped in the past* is a dichotomous variable (1=helped in past and 0=had not helped in past).

Types of Job Help That Resulted in a Job Offer

We use a series of four dichotomous variables to measure the type(s) of help contacts provided job seekers that may have resulted in job seekers receiving job offers. We focused on the job that contacts gave job seekers the most help with and recorded the types of help contacts provided. These types of help include: (1) the sharing of information about job openings, (2) agreeing to be a reference or formal recommendation, (3) the sharing of information about a job or employer (e.g., the pay of the job or the culture at a particular work organization), or 4) the provision of two or more of these types of help.[10] Each variable is coded one if contacts provided the specific kind of help and zero if they did not.

Analytic Sample, Weights, and Missing Data

We begin our analysis with a sample of 564 SNEAS respondents.[11] When we discuss the factors that affect whether contacts help with job searches, we exclude 113 people who were not recently asked to provide assistance with a job search, 49 contacts who were not asked to provide any of the four types of job-related help asked about on the questionnaire, and 2 contacts who had missing values for the variables measuring whether they provided help to job seekers. After making these deletions, we are left with a sample of 400 contacts.

[10] Recall that the SNEAS questionnaire asked respondents whether job seekers requested four types of job-related help: information about a job opportunity, information about a particular job or employer, a reference or letter of recommendation, and an introduction to someone who could help with their search. The SNEAS questionnaire did not ask respondents whether any of the introductions they provided resulted in job seekers applying for or being offered jobs. As a result, we can only examine whether three types of help result in job offers here.

[11] Of the 577 original SNEAS respondents, 13 were missing data for the variables we used to construct the weights and so they are excluded from the sample.

Table 1. Descriptive statistics

	Mean	St. Deviation	Minimum	Maximum
Asked to help (*n*=564)	0.82	—	0	1
Provided help (*n*=400)	0.85	—	0	1
Help resulted in offer (*n*=188)	0.56	—	0	1
Respondent characteristics				
Female (*n*=564)	0.50	—	0	1
Age (*n*=564)				
34 years or younger	0.31	—	0	1
35-44 years	0.18	—	0	1
45-54 years	0.19	—	0	1
55-64 years	0.16	—	0	1
65 years or older	0.16	—	0	1
Minority (*n*=552)	0.18	—	0	1
Employed (*n*=551)	0.77	—	0	1
Contact characteristics				
Female (*n*=400)	0.50	—	0	1
Age (*n*=400)				
34 years or younger	0.07	—	0	1
35-44 years	0.26	—	0	1
45-54 years	0.20	—	0	1
55-64 years	0.20	—	0	1
65 years or older	0.26	—	0	1
Minority (*n*=390)	0.21	—	0	1
Employed (*n*=394)	0.76	—	0	1
Education (*n*=395)	3.37	1.40	0	5
Tie strength				
Closeness (*n*=393)	2.00	0.81	1	3
Familiarity (*n*=396)	2.21	0.77	1	3
Years known (*n*=388)	3.04	1.08	1	3
Helped by job seeker in past (*n*=394)	0.65	—	0	1

Source: Social Networks and Employment Assistance Survey.
Analyses are weighted for respondents' urban/rural status, gender, and age.
Descriptive statistics were calculated before multiple imputation and so sample size varies.

When we examine the characteristics of contacts whose assistance leads to job offers, we exclude 72 contacts who did not help job seekers, 35 contacts whose help did not result in the job seeker applying for the job, and 105 contacts who did not know whether the help they provided lead to a job offer. This leaves us with a sample of 188 contacts.[12] Finally, to

[12] In analysis not shown, we estimated a weighted means comparison test to identify whether contacts who knew whether job seekers applied for or were offered the job they helped them with were different than contacts who did not know this information. Not surprisingly, contacts who knew whether job seekers applied for jobs reported feeling closer to job seekers ($p<.05$), were more likely to have received a favor in the past from job seekers ($p<.05$), were more familiar with job seekers' work qualifications ($p<.10$), and felt job seekers were more trustworthy ($p<.10$) than contacts who did not know this information. These findings suggest the presence of stronger job seeker-contact ties when a contact knows information about the job seekers' networking attempts than when the contact does not know such information. Excluding from analyses, as we

examine the types of help that are most likely to result in job offers, we identified the job that job seekers were trying to get with contacts' help. For job seekers who were trying to get multiple jobs with contacts help, we focused on the "main job" job seekers were trying to get—that is, the job that they asked for the most help in getting. We were unable to determine the "main job" contacts were helping job seekers with for 21 contacts and so they are excluded from our final analysis.

We weight all analyses to adjust for the overrepresentation of rural Washington state residents, women and older respondents in the SNEAS dataset, and use multiple imputation to handle missing data.

RESULTS

How Frequently are People Asked to Help with Others' Searches?

Descriptive statistics (Table 1) reveal that the majority of respondents (just over 80 percent) had recently been asked to help with someone's job search.

Table 2. Logistic regression estimates predicting being asked to help with a job search (n=564)

	β	$\exp^{\beta}$
Respondent's characteristics		
Female respondent	-0.10 (.24)	0.91
Age (65 years or older omitted)		
34 years or younger	1.13** (.39)	3.10
35-44 years	1.30** (.45)	3.67
45-54 years	1.28*** (.37)	3.58
55-64 years	0.74* (.32)	2.10
Minority respondent	1.27* (.55)	3.54
Respondent is employed	1.10* (.26)	3.00
Constant	0.02	

Source: Social Networks and Employment Assistance Survey.
Standard errors are in parentheses.
Analyses are weighted for respondents' urban/rural status, gender, and age.
†p<.10,*p<.05, **p<.01, ***p<.001; two-tailed test.

do, the contacts and job seekers with "weak ties" may affect findings. For example, if job seekers are more likely to be offered jobs when they search using a weak tie (and weak ties are underrepresented in our analysis), our estimates of job offers will be downwardly biased. Future research should seek data from both contacts and job seekers for more accurate data on the outcomes of social networking.

Who is Asked to Help?

To determine the kinds of people who are likely to be asked to help with a job search, we estimated a logistic regression model predicting whether respondents were asked for help. We present the findings from this analysis in Table 2.

Female respondents are just as likely as male respondents to have recently been asked for help with a job search. Some—namely those under age 65, racial/ethnic minorities, and the employed—are more likely to be targeted for a job search than others—individuals 65 years and older, white, and the unemployed. The odds that people under the age of 65 are asked to help with a job search are significantly greater than the odds of 65 year olds and older, holding constant all other demographic characteristics. Racial/ethnic minorities are more likely to be asked to help with a job search than whites—the predicted odds of them being asked to help are 3.5 times the odds of whites ($p<.05$), net of controls. Finally, the odds that employed respondents were recently asked for help are about 3 times the odds of unemployed people ($p<.001$). In sum, contacts' social characteristics differ from non-contacts in important ways.

Who Helps with a Job Search?

Next, we estimated a logistic regression model predicting contacts' provision of help with a job search to determine whether some contacts are more likely to help than others (see Table 3).

Three characteristics of contacts and one indicator of tie strength significantly predict contacts' provision of help. First, employed contacts are more likely to help than unemployed contacts. In particular, the odds that employed contacts help are about 3.75 times the odds of unemployed contacts ($p<.01$). At the same time, when asked, male contacts help more than female contacts—the odds that female contacts help are about 55 percent the odds of men, net of controls ($p<.10$). We also find that contacts who are between the ages of 45 and 54 are less likely to help than contacts who are 65 years or older—the odds that 45-54 year olds help are 62 percent the odds of people who are 65 or older ($p<.05$). Finally, contacts' familiarity with job seekers' work qualifications affects contacts' provision of help with a job search. A one-unit increase in a contact's familiarity with a job seeker's work qualifications results in a 135 percent increase in the odds of helping with their search ($p<.001$). Contacts' level of education and race do not affect their likelihood of helping others with job searches, nor does the length of time contacts have known job seekers, how close they feel to job seekers, or whether they received help from job seekers in the past. In short, attributes related to the contact-job seeker relationship play a minimal role in determining who provides job search help.

Who Provides Help That Results in a Job Offer?

Table 4 shows the findings from a logistic regression model predicting contacts' provision of help that results in a job offer.

Table 3. Logistic regression estimates predicting helping with a job search (*n*=400)

	β	exp^{β}
Contact's characteristics		
Employed	1.33*** (.38)	3.78
Education	0.001 (.13)	1.00
Female	-0.78* (.39)	0.46
Minority	0.10 (.44)	1.11
Age (65 years or older omitted)		
34 years or younger	0.13 (.80)	1.14
35-44 years	0.29 (.58)	1.34
45-54 years	-0.97* (.45)	0.38
55-64 years	-0.27 (.45)	0.76
Strength of relationship		
Years known	-0.09 (.17)	0.92
Closeness	-0.16 (.29)	0.85
Familiarity	0.85*** (.24)	2.35
Helped by job seeker in past	0.53 (.42)	1.71
Constant	-0.06	

Source: Social Networks and Employment Assistance Survey.
Standard errors are in parentheses.
Analyses are weighted for contacts' urban/rural status, gender, and age.
†p<.10,*p<.05, **p<.01, ***p<.001; two-tailed test

Minority and female contacts and contacts who had received favors from job seekers in the past are less likely to provide help that results in job offers than white, male contacts or contacts who had not received favors from job seekers in the past. The odds that minority contacts help job seekers in ways that result in job offers are 84 percent the odds of white contacts (p<.01), while the odds that female contacts help job seekers receive offers are 48 percent the odds of men, although this finding is marginally statistically significant (p<.10). The odds that contacts who received favors help job seekers receive offers are 65 percent the odds of contacts who did not receive favors (p<.10). On the other hand, contacts familiar with job seekers' work qualifications are more likely to provide help that results in job offers. With every one-unit increase in contacts' familiarity with job seekers, the odds that contacts' help results in a job offer increase by 177 percent. Contacts' age, gender, employment status, and education level do not affect their likelihood of providing help that results in a job offer. The two remaining measures of tie strength—closeness and years known—also do not predict contacts' provision of help that results in a job offer.

Table 4. Logistic regression estimates for job seekers receiving a job offer (*n*=181)

	β	**exp^β**
Contact's characteristics		
Employed	0.22 (.50)	1.02
Education	-0.01 (.16)	0.99
Female	-0.65† (.39)	0.52
Minority	-1.81** (.59)	0.16
Age (65 years or older omitted)		
34 years or younger	0.02 (.79)	1.02
35-44 years	0.82 (.57)	2.25
45-54 years	0.45 (.67)	1.56
55-64 years	0.40 (.54)	1.49
Strength of relationship		
Years known	-0.34 (.22)	0.71
Closeness	0.06 (.33)	1.07
Familiarity	1.02*** (.35)	2.77
Helped by job seeker	-1.05† (.61)	0.35
Constant	-0.28	

Source: Social Networks and Employment Assistance Survey.
Standard errors are in parentheses.
Analyses are weighted for contacts' urban/rural status, gender, and age.
†p<.10,*p<.05, **p<.01, ***p<.001; two-tailed test.

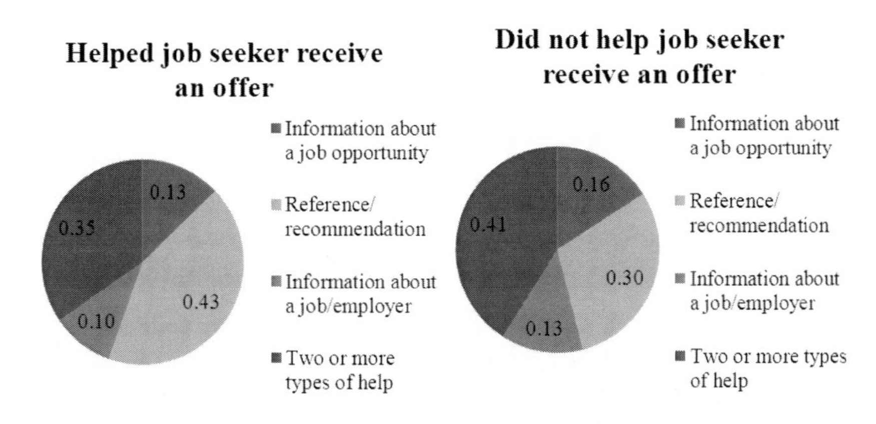

Figure 2. Comparison of the help contacts provided that did and did not result in job offers (*n*=165).

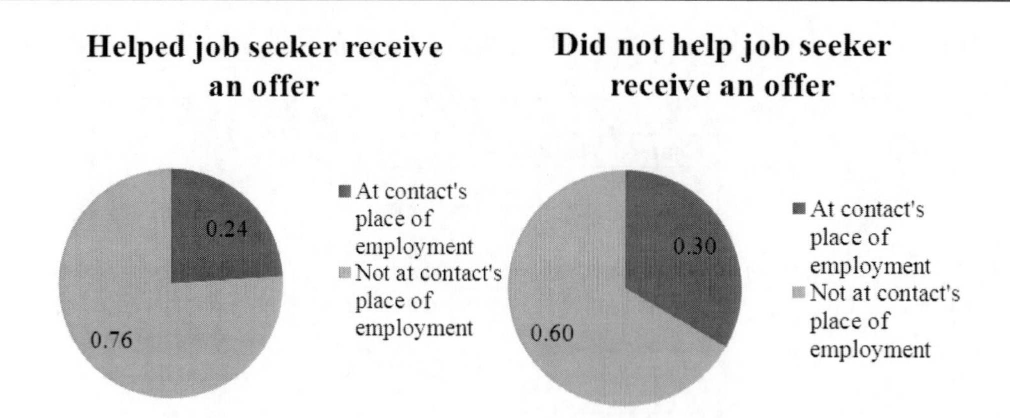

Figure 3. Contacts' employment location by receipt of job offer (n=165).

Is the Type of Contact Assistance Provided Associated with the Chances of Receiving a Job Offer?

To address the specific role contacts have in job seekers' success with their searches, we compared the types of help contacts provide job seekers who receive job offers with the types of help contacts give job seekers who do not receive offers. Figure 2 shows the help contacts provided when job seekers were and were not offered jobs.

Surprisingly, we do not find statistically significant differences in the types of help contacts provide to job seekers who do and do not receive offers. One of the most notable differences—although not statistically significant—is that contacts who help job seekers receive offers provide more references or letters of recommendation (43 percent) than contacts who do not help job seekers receive offers (30 percent). We interpret these findings with caution, however, due to the small sample size; just 89 contacts provided help that resulted in a job while 76 contacts did not.

Figure 3 shows the share of contacts who are employed at the work organization where the job seekers are attempting to find a job by whether or not contact assistance resulted in a job offer.

Although contacts who did not help job seekers receive offers were, on average, more frequently employed at the same organization than contacts that did help job seekers receive offers, this difference is not statistically significant.

Discussion

The purpose of this chapter was to shed much needed light on social network contacts. Contacts play an integral role in the job search—without the timely and valuable resources they provide, job seekers could not find work with their help or experience the myriad of benefits that come with using this job search strategy (see Marsden and Gorman 2001; Ioannides and Loury 2004). Scholars have a limited understanding of contacts' role in the job search because previous research has tended to rely on methodological approaches that downplay their importance or overlook their influence all together (but see Marin 2012;

Smith 2005, 2007, 2010). In this chapter, we argued that scholars should focus more of their energy on contacts, in addition to pursuing studies of job seekers who use their networks to search for work and organizations who recruit through their employees' personal networks.

We presented data from a unique survey which asked people about the last time they may have been asked to help with someone's job search. Our findings from the Washington state sample corroborate what others have found (Granovetter 1995; Marsden and Gorman 2001); social networking is a common occurence—82 percent of SNEAS respondents were recently asked by someone looking for work to help with their search. Analyses also show that most contacts (85 percent) provided the help job seekers needed to advance their search, and when contacts do help, about half of contacts reported that the help they provided resulted in job seekers receiving job offers.

Despite the fact that being asked to help with a job search is common, our analysis shows that some people are more likely than others to be asked. Job seekers shy away from asking people over the age of 65 for help with their job searches. This effect holds after controlling for employment status which suggests that job seekers are not turning to younger people for help with their job searches because they are more likely to be attached to the labor force than people over age 65. Recent research by McDonald and Mair (2010) may help explain why older individuals are less likely to be asked to help with a job search; daily interactions with members of individuals' networks decrease with age. Perhaps older individuals are less likely to be asked for help with a job search than younger because they are not likely to have the types of social interactions in which networking occurs.

We also find that racial/ethnic minorities are more likely to be asked to help with a job search than whites. This finding is expected given previous research which shows that racial/ethnic minorities search for work using this strategy more than whites (Corcoran, Datcher and Duncan 1980; Falcón 1995; Falcón and Melendez 2001) and that minority job seekers tend to find work through same-race/ethnic contacts (Falcón and Melendez 2001; Stainback 2008). If minority job seekers frequently search for work using minority contacts, then it stands to reason that minorities are likely to be sought out for help with a job search more than whites.

Finally, our analyses show that employed contacts are more likely than unemployed contacts to have recently been asked to help with a job search. This finding suggests that job seekers turn to contacts with ties to the labor market for help with their job search, possibly because they expect employed contacts will be in a better position than unemployed contacts to help them find work. Although we are unable to pinpoint the reasons job seekers turn to particular contacts for help with their job searches, all in all, these findings suggest that job seekers' networking choices are shaped by two factors: access to particular types of contacts (i.e., younger or minority contacts), and contacts' likelihood of providing the kinds of resources that would be beneficial for a job search. Future research should test these possibilities by examining how job seekers come to decide who they will ask for help.

Our second research question examined whether some job seekers were more likely to provide help than others and the factors that motivated this help. We find that employed, male contacts familiar with job seekers' work qualifications are more likely to help job seekers than unemployed, female, or unfamiliar contacts. Employed contacts are likely to be in a position to help with job searches because they may be more likely than the unemployed (particularly, the unemployed who are not looking for work) to routinely learn of job opportunities, have direct access to employers making hiring decision, and to maintain their own work-related connections which they can draw on to help others. An alternative explanation is that unemployed contacts may hesitate to help because they are interested in job opportunities for themselves and do not want to increase their competition by sharing job opportunities with others. Future research should examine employed and unemployed contacts' knowledge of job leads and work-related connections to distinguish between these possible explanations.

We also find that women are less likely to help with job searches than men. One explanation for this finding is that women are limited in the job-related resources they possess. Due to workplace sex discrimination and occupational sex segregation, women may occupy less powerful positions than men and so they are less able to help when asked. Women continue to be underrepresented in many supervisory and management positions (Stainback and Tomaskovic-Devey 2009) and so they miss out on hearing about job opportunities or lack access to the information that could give job seekers a competitive advantage. For example, an administrative assistant is less likely than her employer to know when her work organization will be experiencing a labor demand and planning to hire. Women are also less likely than men to have high status workplace connections (McGuire 2000) and so they may be less able to exert their own influence on behalf of job seekers.

Finally, we find that contacts are more likely to help job seekers whose work qualifications they are familiar with. This finding is in line with Marin's (2012) work which shows that contacts share job leads with strong ties more frequently than weak ties because they are certain that job seekers will be interested in and appreciative of job leads. Taken together, our and Marin's (2012) findings suggest that familiar contacts help because they know the help is wanted and they know enough about the job seeker's qualifications to be able to help.

We also examined the factors that affected whether contacts' help leads to a job offer. Contacts who are white, male, familiar with job seekers' work qualifications, and who have not received favors from job seekers in the past provide the kind of help that lands job seekers job offers. These findings suggest that contacts' characteristics and the relationships contacts have with job seekers are important predictors of job seekers receiving offers, maybe even more important than job seekers' characteristics. For example, job seekers who search for work using white contacts are more likely to be offered jobs than job seekers who use minority contacts, but job seekers' race does not affect their likelihood of receiving a job offer.

One explanation for this finding could be due to white men's superior labor market status relative to minorities. Although we control for contacts' employment status in models, we do not account for occupation or job prestige. Since white men tend occupy jobs with more power and authority than minorities (Elliott and Smith 2004; Smith 2002; Tomaskovic-Devey 1993) and minorities are concentrated in jobs that pay less than jobs dominated by whites (Catanzarite 2003; Kmec 2003), these findings may reflect whites' better access to high quality job information. Moreover, in line with Smith's (2005, 2007, 2010) work, minorities' limited opportunities at work may affect the quality of help they provide because they are afraid of tarnishing their workplace reputations. Having limited power and authority may result in minorities being guarded about "how much" they help others. Thus, the lower quality help minority contacts may provide job seekers would seemingly decrease job seekers' chances of being offered jobs relative to job seekers who use white contacts. A study which focuses on the quality of help contacts provide and how this varies by contacts' race would help clarify how job seekers chances of being offered jobs diminish when they use minorities to search for work compared with white men.

A final explanation for job seekers' diminished chances of being offered jobs when using minority contacts compared with white contacts has to do with the returns on the help received (Lin 2001). Previous research has demonstrated that returns on social capital depend on contacts' race. Kmec and Trimble's (2009) research on the effects of social network contact use in the job attainment process on pay demonstrate this possibility. Kmec and Trimble (2009) studied how workers' pay was affected by contacts' race, the type of help contacts provided to them, and whether contacts were employed by same employer as them. They found that workers who used black contacts to find their jobs earned better pay than workers who used formal methods, but only when their contact's race was unknown to the potential employer. For example, when black contacts told workers about jobs and when they were not employed by the same employer, workers experienced a pay premium. Future research should focus on employers' reactions to referrals and the other types of help contacts provide (e.g., when contacts talk with employers on job seekers behalf, etc.) to see whether employers value it differently based on contact race/ethnicity.

We find that job seekers who ask for help from contacts who are familiar with their work qualifications and who have not received favors from them in the past are more likely to receive job offers than job seekers who ask for help from contacts who are unfamiliar with them or who have received favors in the past. How does contacts' familiarity with job seekers result in job seekers' increased chances of being offered a job? Fernandez, Castilla and Moore (2000) suggested a mechanism that helps explain this finding. Fernandez and his colleagues argued that job seekers who find work through contacts are "better matched" to their jobs than job seekers who find work through formal methods. This is because contacts have superior job information about job seekers and employers, and they use this information to refer workers who they know will be good fits. Perhaps contacts' familiarity with job seekers' work qualifications affects the extent to which contacts can pass along job information and influence that is a "good fit" for job seekers. In other words, they use their familiarity of job seekers to pass along the social resources that are most likely to result in job offers.

Our analysis also shows that contacts who received a favor from a job seeker in the past, whether personal or professional, are less likely to provide help that results in a job offer. This finding is surprising because research on social exchanges suggests that a history of exchanging favors in the past signals that the job seeker will continue to provide favors in the future (Smith 2007). In other words, the more exchanges contacts have with job seekers, the more willing they should be to provide "good quality" help—the the kind of help that results in job offers—because they know they are likely to receive help from job seekers in the future. Data limitations prohibit us from knowing whether contacts and job seekers were in the habit of exchanging favors, or if contacts had only received one favor from job seekers in the past. Perhaps some contacts received "one-time" help from job seekers and so they felt obligated to reciprocate in order to "save face," but did not feel like they needed to provide quality help. Another possibility is that contacts feel obligated to help because they have received favors from job seekers, but are not able to provide quality help. In other words, they agree to help because they feel like they have to, not because they can provide the best possible resources for the job seeker.

Finally, our fifth research question explored whether contacts who helped job seekers receive offers provided different help than contacts who did not help job seekers receive offers. To answer this question, we examined the types of help contacts provided, as well as whether contacts were employed at the organizations that job seekers were trying to find employment. We found no difference in the types or prevalence of help provided by contacts whose help led to a job offer and those whose help did not lead to a job offer. It may be that contacts that provide assistance are not always told whether their help leads to a job offer or, just as plausible, that they do not remember this information. For this reason, we may be undercounting the extent to which contact assistance leads to actual job offers. To prevent this potential recall error, future research should inquire about more recent job assistance (i.e., that provided in the last 3 months). If possible, researchers may also want to ask the name and contact information of the person who received the assistance and do a follow-up survey to identify if they received the job offer their contact mentioned.

Our analyses demonstrate that directly studying contacts can provide scholars with insights into the mechanisms that make social network use a successful search strategy for some, but not others. Job seekers who network with the unemployed, women, or contacts that are unfamiliar with their job qualifications may receive a low return on their networking investment because these contacts are less likely to help than employed, male, or familiar contacts. Some groups of job seekers may be more likely to turn to unemployed, female, or unfamiliar contacts than others and so they are systematically getting less return on their networking investments. For example, poor, urban minorities are more likely to have unemployed people in their social networks than the well-off or whites (Tigges, Browne, and Green 1998) and so this finding illustrates how networking may be less effective for them and contributing to long-term unemployment among this group. Previous research has also demonstrated that men tend to find jobs with the help of other men, and women with the help of other women (Hanson and Pratt 1991; Mencken and Winfield 2000; Straits 1998). To the extent that female job seekers systematically turn to other women when networking, our analyses show how networking can be less advantageous for women than men.

Finally, we find that white, male and familiar contacts who had not received favors from job seekers in the past are more likely to lead job seekers to employment offers than minority, female, or unfamiliar contacts who had received favors from job seekers in the past. These findings demonstrate another way in which contact use is more effective for some job seekers than others: some contacts are better at helping job seekers become employed than others. For example, because minority job seekers are more likely to network with minority contacts than white contacts (Falcón and Melendez 2001; Stainback 2008), our findings show how minority job seekers do not get as much "bang for their buck" as white job seekers do when networking and may help explain persistent unemployment among some minority groups.

We are not concluding that job seekers avoid asking female or minority contacts for job assistance because women and minorities are doing something "wrong" per se. These job seekers may not have gotten a job offer had they relied on a male or white contact instead—we do not have the data to test this possibility. It may be that employers respond differently to the assistance provided by white men (i.e., take it more seriously, follow through on it) than by women and minorities and this explains, in part, the differences we find between white men, women and minorities. To test this possible explanation requires data from employers who use referrals in the hiring process.

CONCLUSION

The goal of this chapter was to use an under-utilized methodological approach (i.e., studying contacts rather than job seekers or employers) to shed light on the social networking process as it applies to job attainment. This approach builds upon previous research on social network contact use during the job search by highlighting potential mechanisms driving social network contact use. This research is not without limitations though. First, findings gathered from this study are only generalizable to adults living in Washington state and Washington state differs from other U.S. states in ways that may affect the social networking process, in particular the employment outcome of those who use networks in the job attainment process. For example, Washington guarantees more equal employment protections (e.g., protections based on sexual orientation, gender identity, an equal pay law that expands the federal government law) than a majority of states (Kmec and Skaggs 2012). Such protections may mean employers in the state abide by more formal hiring processes and so network ties matter less in the employment process. It also has one of the highest state minimum wages, a factor that could affect network tie reliance in the opposite direction. Higher minimum wages equate into higher turnover costs so employers might rely more heavily on contact referrals that can help them better match applicants to jobs. Future research should strive to administer a similar survey to a nationally representative sample of individuals.

The survey data we analyze was collected during the recent Great Recession (2009-2010) and it is unclear how the economic downturn may have affected our results. On the one hand, the recession may have increased contacts' likelihood of providing help because contacts realized that many job seekers were desperate to find work. On the other hand, contacts may have been more reluctant to help than usual in case they needed to use their own social resources (for example, in the event that they lost their job). Job seekers' success with social networking was undoubtedly affected by the Great Recession—regardless of how job seekers looked for work, their likelihood of being offered jobs was lower than usual because employers were not hiring during this time period. A study during a more economically stable time would reveal how economic conditions affect contacts' role in the social networking process.

Despite these limitations, the data we have analyzed are unique; they are among the first to examine the job seeking process from the perspective of someone asked to provide help with job attainment. For this reason, the analyses presented here provide a template for future studies. We envision a number of ways researchers can build on the approach we have laid out here. To begin, researchers should attempt to study social network contact use during the job search from the perspectives of *all* parties involved. Finding work using social network contacts involves three key players: the job seeker, the contact, and the employer. The story we tell in this chapter is one-sided—for a reason—because we exclusively study contacts. A study that "matches" data from a job seeker and contact would provide a more complete picture of the networking process and address several of the limitations of this research. For example, we found that some contacts are more likely to be the targeted for help with a job search than others. We speculate that job seekers are "strategic" in choosing social network contacts, but because we focused on contacts rather than job seekers, we were unable to determine why job seekers selected particular social network members for help. By collecting "matched" data from a job seeker and a contact, researchers can understand how job seekers make decisions about the people they turn to during their job search and how job seekers' strategies in choosing social network contacts affects contacts' responses to their requests for help. A "matched" data set would also provide more complete information about particular instances of social networking. For example, our analysis was limited by the fact that many contacts did not know whether job seekers applied or were offered jobs as a result of the help they provided. A study which surveys both job seekers and contacts has the potential to provide more complete information on a particular networking act.

Future research should pursue creative ways of collecting data not only from job seekers and the contacts they ask for help, but also from potential employers. This is particularly important for understanding how social network contact use affects job seekers' likelihood of becoming employed. Our research does not consider demand-side processes that may affect whether job seekers succeed when they use their social network contacts. Employers vary in the degree to which they "value" referrals (Neckerman and Fernandez 2003)—that is, some employers prefer "networked" job seekers while others do not. Regardless of the advantages social network contacts provide job seekers, some job seekers will be not benefit from social network contact use simply because employers do not favor "networked" job seekers or because some companies either limit the extent to which referrals can matter (e.g., in a highly formalized system with third-party oversight of the employment process versus when employers offer referral bonuses to employees). A data set that matches information from job seekers, contacts, and potential employers would allow researchers to account for these possibilities.

As social networking during the job process becomes more prevalent with the growth of internet-based networking sites and the way that the internet has made networking less bound by geography, researchers stand to gain by studying all aspects of the job search, in particular the contact who is the essential bridge between the employer and job seeker.

REFERENCES

Allison, P. (2001). *Missing Data*. Thousand Oaks, CA: Sage Publications.

American Association for Public Opinion Research (AAPOR). (2010). *Standard Definitions: Final Dispositions of Case Codes and Outcome Rates for Surveys*. Retrieved November 15, 2010 (http://www.aapor.org/Standard_Definitions1.htm).

American Community Survey (ACS). (2009). 2009 American Community Survey 1 Year Estimates: Washington: Detailed Tables. Generated by Lindsey B. Trimble using American Factfinder, U.S. Census Bureau, U.S. Dept. of Commerce. Retrieved December 3, 2010 (http://factfinder2.census.gov/faces/nav/jsf/pages/index.xhtml).

Battaglia, M. P., Link, M. W., , Frankel, M.R., Osborn, L. & Mokdad, A.. (2008). An Evaluation of Respondent Selection Methods for Household Mail Surveys. *Public Opinion Quarterly*, *72*, 459-469.

Beggs, J. J. & Hurlbert, J. S. (1997). The Social Context of Men's and Women's Job Search Outcomes. *Sociological Perspectives*, *40,* 601-622.

Bian, Y. (1997). Bringing Strong Ties Back in: Indirect Ties, Network Bridges, and Job Searches in China. *American Sociological Review*, *62*, 366-385.

Burt, R. (1992). *Structural Holes: The Social Structure of Competition*. Harvard University Press, Cambridge, Mass.

------. (2001). Bandwidth and Echo: Trust, Information, and Gossip in Social Networks. In J. E. Rauch and A. Casella (Eds.) *Networks and Markets* (pp. 30-74). New York: Russell Sage Foundation,.

Catanzarite, L. (2003). Race-Gender Composition and Occupational Pay Degradation. *Social Problems, 50*, 14-37.

Chapple, K. (2006). Overcoming Mismatch: Beyond Dispersal, Mobility, and Development Strategies. *Journal of the American Planning Association, 72*, 322-336.

Corcoran, M, Datcher, L. & Duncan, G.. (1980). Information and Influence Networks in Labor Markets. In G. J. Duncan and J. N. Morgan (Eds.) *Five Thousand American Families—Patterns of Economic Progress*, vol. *VIII* (pp. 1-37). Ann Arbor: Institute for Social Research.

Dillman, D. A., Smyth, J. D., & Christian, L.M. (2009). *Internet, Mail, and Mixed-Mode Surveys: The Tailored Design Method.* Hoboken, New Jersey: John Wiley & Sons, Inc..

Elliott, J. R. Referral Hiring and Ethnically Homogeneous Jobs: How Prevalent is the Connection and for Whom? *Social Science Research, 30*, 401-425.

Elliott, J. R. & Smith, R. A. (2004). Race, Gender, and Workplace Power. *American Sociological Review, 69*, 365-386.

Falcón, L. M. (1995). Social Networks and Employment for Latinos, Blacks, and Whites. *New England Journal of Public Policy, 11*, 17-28.

Falcón, L. M., & Melendez, E.. (2001). The Social Context of Job Searching for Racial Groups in Urban Centers. In A. O'Connor, C. Tilly, and L. Bobo (Eds.) *Urban Inequality: Evidence from Four Cities* (pp. 341-371) . New York: Russell Sage.

Fernandez, R. M., & Mors, M. L.. (2008). Competing for Jobs: Labor Queues and Gender Sorting in the Hiring Process. *Social Science Research, 37*, 1061-1080.

Fernandez, R. M., & Fernandez-Mateo, I. (2006). Networks, Race, and Hiring. *American Sociological Review, 71*, 42-71.

Fernandez, R. M., & Sosa, M.L. (2005). Gendering the Job: Networks and Recruitment at a Call Center. *American Journal of Sociology, 111*, 859-904.

Fernandez, Roberto M. & Nancy Weinberg. (1997). Sifting and Sorting: Personal Contacts and Hiring in a Retail Bank. *American Sociological Review, 62*, 883-902.

Fernandez, R. M., Castilla, E. J., Moore, P. (2000). Social Capital at Work: Networks and Employment at a Phone Center. *The American Journal of Sociology, 105*, 1288-1356.

Granovetter, M. (1995). *Getting a Job: A Study of Contacts and Careers (2nd edition).* Chicago: University of Chicago Press.

Green, G. P., Tigges, L. M., & Browne, I.. (1995). Social Resources, Job Search, and Poverty in Atlanta. *Research in Community Sociology, 5*, 161-182.

Hanson, S., & Pratt, G. (1991). Job Search and the Occupational Segregation of Women. *Annals of the Association of American Geographers, 81*, 229-253.

Holzer, H. J. (1996). *What Employers Want: Job Prospects for Less-Educated Workers.* New York: Russell Sage Foundation.

Huffman, M. L., & Torres, L. (2002). It's Not Only "Who You Know" that Matters: Sex, Personal Contacts, and Job Lead Quality. *Sex & Society, 16*, 793-813.

Ioannides, Y. M., & Datcher Loury, L. (2004). Job Information Networks, Neighborhood Effects, and Inequality. *Journal of Economic Literature, 42*, 1056-1093.

Johnson, J. H., Bienenstock, E. J., & Farrell Jr., W. C. (1999). Bridging Social Network and Female Labor-Force Participation in a Multiethnic Metropolis. *Urban Geography*, *20*, 3-30.

Kmec, J. (2003). Minority Job Concentration and Wages. *Social Problems*, *50*, 38-59.

------. (2007). Ties that Bind: Race and Networks in Job Turnover. *Social Problems*, *54*, 483-503.

Kmec, J A., & Skaggs, S. L. (2012). The State of Sex-Based Rights Regulation at Work and Managerial Sex Diversity. Unpublished manuscript.

Kmec, J. A. & Trimble, L. B. (2009). Does it Pay to Have a Network Contact? Social Network Ties, Workplace Racial Context, and Pay Outcomes. *Social Science Research*, *38*, 266-278.

Lin, N. (1999). Social Networks and Status Attainment. *Annual Review of Sociology*, *25*, 467-487.

------. (2000). Inequality in Social Capital. *Contemporary Sociology*, *29*, 785-795.

------. (2001). *Social Capital: A Theory of Social Structure and Action*. New York: Cambridge University Press.

Lin, N., & Ao, D. (2008). *The Invisible Hand of Social Capital: An Exploratory Study*. In N. Lin and B. H. Erickson (Eds.) Social Capital: An International Research Program (pp. 107-132). Oxford: Oxford University Press.

Lin, N., Ensel, Walter M. & Vaugh, J. C. (1981). Social Resources and Strength of Ties: Structural Factors in Occupational Status Attainment. *American Sociological Review*, *46*, 393-405.

Marin, A. (2012). Don't Mention It: Why People Don't Share Job Information, When They Do, and Why it Matters. *Social Networks*, *34*, 181-192.

Marsden, P. V. (1996). The Staffing Process: Recruitment and Selection Methods. in . Kalleberg, D. Knoke, P. V. Marsden, & J. L. Spaeth (Eds.) *Organizations in American: Analyzing Their Structures and Human Resource Practices* (pp.133-156).. Thousand Oaks, CA: Sage Publications.

Marsden, P. V., & Gorman, E. H. (2001). "Social Networks, Job Changes, and Recruitment." Pp. 467-502 in *Sourcebook of Labor Markets: Evolving Structures and Processes*, edited by I. Berg & A. Kalleberg, New York: Kluwer Academic.

Marsden, P. V., & Hurlbert, J. S. (1988). Social Resources and Mobility Outcomes: A Replication and Extension. *Social Forces*, *66*, 1038-1059.

McDonald, S. (2011). What's in the Old Boys' Network? Accessing Social Capital in Gendered and Racialized Networks. *Social Networks*, *33*, 317-330.

McDonald, S., & Day, J. (2010). Race, Gender, and the Invisible Hand of Social Capital. *Sociology Compass*, *4*, 532-543.

McDonald, S., & Elder Jr., G. H. (2006). When Does Social Capital Matter? Non-searching for Jobs across the Life Course. *Social Forces*, *85*, 521-549.

McDonald, S., & Mair, C. A. (2010). Social Capital Across the Life Course: Age and Gendered Patterns of Network Resources. *Sociological Forum*, *25*, 335-359.

McDonald, S., Lin, N., Ao, D. (2009). Networks of Opportunity: Gender, Race, and Job Leads. *Social Problems*, *56*, 385-402.

McGuire, G. (2000). Gender, Race, Ethnicity, and Networks: The Factors Affecting the Status of Employees' Network Members. *Work and Occupations*, *27*, 501-523.

Mencken, C., F., & Winfield, I. (2000). Job Search and Sex Segregation: Does Sex of Social Contact Matter? *Sex Roles*, *42*, 847-864.

Miller, S. R., & Rosenbaum, J. E. (1997). Hiring in a Hobbesian World: Social Infrastructure and Employers' Use of Information. *Work and Occupations*, *24*, 498-523.

Mouw, T. (2003). Social Capital and Finding a Job: Do Contacts Matter? *American Sociological Review*, *68*, 868-898.

Neckerman, K. & Fernandez, R. M. (2003). Keeping a Job: Network Hiring and Turnover in a Retail Bank. *Research in the Sociology of Organizations*, *20*, 299-318.

Newman, K. (1999). *No Shame in My Game: The Working Poor in the Inner City*. New York: Knopf and the Russell Sage Foundation.

Petersen, T., Saporta, I., & Seidel, M.D.L. (2000). Offering a Job: Meritocracy and Social Networks. *American Journal of Sociology*, *106*, 763-816.

Podolny, J. M. (2005). *Status Signals*. Princeton: Princeton University Press.

Powell, W. & Smith-Doerr, L. (1994). Networks in Economic Life. 379-402 in Neil Smelser and Richard Swedberg (Eds.), *Handbook of Economic Sociology* (pp. 379-402). Princeton, NJ: Princeton University Press.

Royster, D. A. (2003). *Race and the Invisible Hand: How White Networks Exclude Black Men from Blue-Collar Jobs*. Berkeley, CA: University of California Press.

Smith, R. A. (2002). Race, Gender, and Authority in the Workplace: Theory and Research. *Annual Review of Sociology*, *28*, 509-542.

Smith, S. S. (2000). Mobilizing Social Resources: Race, Ethnic, and Gender Differences in Social Capital and Persisting Wage Inequalities. *The Sociological Quarterly*, *41*, 509-537.

------. (2005). Don't Put My Name On It: Social Capital Activation and Job-Finding Assistance Among the Back Urban Poor. *American Journal of Sociology*, *111*, 1-57.

------. (2007). *Lone Pursuit: Distrust and Defensive Individualism Among the Black Poor*. Russell Sage Foundation, New York.

------. (2010). A Test of Sincerity: How Black and Latino Service Workers Make Decisions about Making Referrals. *The ANNALS of the American Academy of Political and Social Science*, *629*, 30-52.

Stainback, K. (2008). Social Contacts and Race/Ethnic Job Matching. *Social Forces*, *87*, 857-886.

Stainback, K., & Tomaskovic-Devey, D. (2009). Intersections of Power and Privilege: Long-Term Trends in Managerial Representation. *American Sociological Review*, *74*, 800-820.

Straits, B. C. (1998). Occupational Sex Segregation: The Role of Personal Ties. *Journal of Vocational Behavior*, *52*, 191-207.

Tigges, L. M., Browne, I., & Green, G. P. (1998). Social Isolation of the Urban Poor. *The Sociological Quarterly*, *39*, 53-77.

Tomaskovic-Devey, D. (1993). *Gender and Racial Inequality at Work: The Sources and Consequences of Job Segregation*. Ithaca, NY: ILR Press.

Trimble, L. B. & Kmec, J. A. (2011). The Role of Social Networks in the Job Attainment Process. *Sociology Compass*, *5*, 165-178.

Williams, C., Muller, C., & Kilanski, K. (2012). Gendered Organizations in the New Economy. *Gender & Society*, *26*, 549-573.

Yakubovich, V. (2005). Weak Ties, Information, and Influence: How Workers Find Jobs in a Local Russian Labor Market. *American Sociological Review*, *70*, 408-421.

In: Social Networking
Editors: X. M. Tu, A. M. White and N. Lu
ISBN: 978-1-62808-529-7

Chapter 10

IMPLICATIONS OF SOCIAL NETWORK ENDOGENEITY: FROM STATISTICAL TO CAUSAL INFERENCES

N. Lu[1], A. M. White[2], P. Wu[1], H. He[1], J. Hu[1], C. Feng[1] and X. M. Tu[1,2]

[1]Department of Biostatistics and Computational Biology
[2]Department of Psychiatry, University of Rochester, Rochester, NY, US

ABSTRACT

A broad spectrum of disciplines have adopted social network data to examine relevant contextual issues in a wide array of fields. Yet, methods to provide valid inference within the confines of social network endogeneity are woefully lacking. Although the concept of endogeneity is not new and methods exist to address this issue in conventional statistical analysis, none of the available approaches provide valid inference when applied to social network data, since the social network engeneity is a completely different phenomenon. This chapter reviews the concept of endogeneity in standard statistical analysis and contrasts this classic concept with the new social network endogeneity to show why conventional statistical models are fundamentally flawed when used to model endogenous relationships within the social network. By leveraging U-statistics, functional response models and potential outcomes, we discuss a new line of approach to address the social network endogeneity for both valid statistical and causal inferences.

Keywords: clustered data, contagion, distribution-free models, functional response models, homophily, marginal structural models, potential outcome, propensity score, social network density, stochastic independence, structural equation models, U-statistics

1. Introduction

Social and behavioral sciences are burgeoning with applications of social networks. At the core of such investigations are questions about social network structures and their impacts on health and behavioral outcomes and community wellbeing, such as adoption of new products (Raghupathi and Fogel, 2013), contagion of disease (Sadilek et al., 2012; Kiuru et al., 2012), facilitation of job search (Trimble et al., 2013), and influence of online social media on social movements such as the Arab Spring (Akpan-Obong and Parmentier, 2013). To investigate such hypotheses related to social network data, conventional statistical models and methods have been applied, such as t-tests, chi-squares and regression models. However, such popular methods are derived from the current predominant paradigm premised on modeling with-subject attributes, they generally yield invalid results when applied to investigating social network structures defined by between-subject attributes. As the corpus of research studies investigating hypotheses related to social network data grows exponentially, it is important to understand this unique endogeneity feature of social network data and develop new approaches to address its impact on inference about social network structures.

In this chapter, we discuss the phenomenon of social network endogeneity induced by the between-subject attributes when modeling social network data. We start with a related concept of endogeneity in conventional data analysis and show why the social network endogeneity is fundamentally different from its counterpart in conventional data, which in particular renders conventional methods to yield invalid conclusions when applied to social network data. We discuss new approaches to address the social network endogeneity. Since a majority of social network applications involve observational studies, we also discuss issues for causal inference with such data.

2. Issues on Statistical Inference

Social network data has been adopted across a broad spectrum of disciplines, and with a wide array of analytic methods, to examine the effects of network contexts on a variety of health, behavioral and community wellbeing outcomes. Yet, statistical methods are lacking to provide valid inference, since all available methods are based on the traditional "subject-centered" paradigm. Relationships in social networks are rather defined by interactions between subjects, and such a between-subject interlocking, or endogenous, relationship introduces an additional dimension that does not fit the confines of the traditional statistical models (Yu et al., 2011; El-Sayed et al., 2012).

2.1. Social Network Endogeneity---A Game-Changing Concept Necessitating a Paradigm Shift

In traditional statistical models, endogeneity refers to a reciprocal relationship between two variables and as a result statistical models for the relationship between such variables may yield invalid inference. For example, when performing liner regression analysis, one variable is designated as the dependent, or response, variable, and other variables that cause changes of this dependent variable are known as independent variables. Unlike the dependent variables, the independent variables can vary freely. In other words, the dependent variable will have any influence on the change of any of the independent variables. If this assumption is violated, regression analysis will yield biased model estimates.

In clinical trial research, dose-response relationships are of great interest when non-compliance is high, because conventional intent-to-treat (ITT) analysis often fails to yield treatment effect in such situations. ITT analysis estimates treatment effect by the randomized treatment groups. Since random treatment assignment is independent of any potential confounder such as disease severity, ITT analysis provides valid inference about treatment difference.

Since often in the real world not everyone complies with treatment, the causal effect estimated by the ITT principle does not really reflect what would happen if the drug was used, since it is the causal effect of telling someone to take the drug verses not to take the drug. Indeed, when compliance is now, an efficacious intervention may fail to show any benefit. In such situations, we are often interested in estimating the "truly treated" or "As treated" effect, i.e., the causal effect of actually taking the intervention, not merely being instructed to do so.

Let x be the actual truth of whether the person took the intervention and y be some continuous outcome of interest. If we simply regression y on x in a linear regression,

$$y = \beta_0 + \beta_1 x + \varepsilon_y, \tag{1}$$

where ε_y denotes the error term for the model, we may get biased estimate of treatment effect. The problem is we may have some confounding variables, such as disease severity, that both influence the likelihood of treatment compliance and cause the change in the outcome regardless of treatment. As a result, x is not really an independent variable, as it correlates with the error term ε_y, causing the breakdown of the regression analysis.

In the causal inference literature, a common approach for addressing such endogeneity is the instrumental variable (Angrist and Krueger, 2001). A variable z is an instrumental variable (IV), if it is direct cause of x and is not a direct cause of y. Within the context of randomized studies, the random treatment assignment z is an IV. If we have an IV variable, we can address endogeneity by regressing y on x and regressing x on z, yielding the following structural equation model (SEM):

$$y = \beta_0 + \beta_1 x + \varepsilon_y, \quad x = \beta_0 + \beta_1 z + \varepsilon_x, \tag{2}$$

where ε_y (ε_x) denotes the error term for the first (second) components of SEM. Although similar in appearance, the SEM is fundamentally different from standard linear regression, since x is neither a dependent nor an independent variable; x is an independent variable in the first and a dependent variable in the second equation. Because of the correlated error terms ε_y and ε_x, the SEM cannot be fit by applying regression analysis to each of the regression components. To estimate the effect of x on y, we must fit the two submodels simultaneously using specialized SEM procures (Bollen, 1989; Kowalski and Tu, 2007; Muthen and Muthen, 2012).

When there exists reciprocal, or feedback, loops between variables, or endogeneity, regression analysis no longer applies and new approaches such as SEM must be used to address the associated inference issue. Endogenous relationships are not only present, but are actually a defining feature of the social network attributes, since social networks are all about relationships between subjects of the social network. Endogeneity among the social network attributes are borne from assessing `social ties' among members or subjects populating the social network systems under investigation. Each subject in the social network forms or has the potential to form reciprocal loops between one or more subjects in the network. Note that we view observed social network relationships in a real study as a sample from a population of social networks. Thus, an absent tie between two subjects does not mean that the two are not potentially connected. Indeed, the probability of connectivity between the two subjects, or density, is actually a primary parameter of interest in social network analysis.

Because of the endogenous relationship between subjects in the social network, standard statistical models such as t-test, chi-square test and regression do not provide valid inference. However, within the current context of social networks, the issue of endogeneity takes on a new meaning, because of a fundamental difference in the type of relationship considered between conventional statistical and social network analysis. Traditional statistical models represent relationships between variables, or measures of individual attributes, such as age and depression. However, social network analysis seeks to model attributes defined based on interactions between subjects such as connections between two subjects, and more generally, relationships between individual attributes and between-subject interactions. For example, in a study on contagion of depression, we are not only interested in relationships between two individuals (between-subject attribute), but also how such relationships influence depressive symptoms for an individual (individual, or within-subject, attribute). The change from modeling within-subject attributes to between-subject dynamics has fundamental implications for valid inference, which in particular precludes applications of any approach premised upon modeling within-subject attributes such as regression and SEM.

The challenge to address endogeneity in the current context of modeling within-subject attributes is the non-independence created by between-subject feedback loops. If we were to address this between-subject endogeneity using methods for traditional statistical models, such as the IV, we would add additional relationships involving other subjects, rather than additional variables. However, as all subjects are potentially connected, this implies that we would include every subject in the model, leaving no "individual data" for estimating the parameters. In other words, the model includes all individual observations in the sample. In contrast, this problem does not arise in the early context of modeling within-subject attributes, since we only add additional variables, or attributes from each individual, in the model and thus the observations in the sample still provides the same amount of independent information for estimating the model parameters, albeit with less accurate estimates due to increased number of parameters.

In addition to adding variables, another major approach for addressing dependence among variables is to partition the sample into independent clusters of observations. A prime example of this approach is found in analysis of clustered data. For example, in a longitudinal study with n subjects, each subject is assessed repeatedly over time and the repeated measures of an outcome of interest are not stochastically independent, because observations are generally more similar within the same individual than across from the different subjects. Again, we will not obtain valid inference, if the dependence among the repeated outcomes within the same individual is ignored. Longitudinal methods partition the sample of observations into n clusters, with each cluster consisting of observations from the same subject. As in the IV approach, the correlated repeated outcomes within the same subject are modeled in longitudinal methods, with the clusters providing the independent information to estimate the model parameters.

This partitioning approach does not work for our setting either. As subjects in a social network are potentially connected, it is not possible to partition the sample into independent clusters. The difficulty again is due to the nature of the between-subject endogeneity in our context. Clustered data is a phenomenon of correlated variables, or within-subject attributes, but social network endogeneity is about interactions between subjects, or between-subject attributes.

Note that in some non- longitudinal clustered data, data clustering may also appear "between-subject" attributes. For example, in a multi-center clinical trial study, subjects within each medical facility are likely to have more similar responses than those across the different sites. Although outcomes from subjects within the facility are correlated, such between-subject correlations are different from the between-subject interaction within our social network context. In particular, we can partition the sample into independent clusters, with the clusters consisting of the different medical facilities, since observations are independent between different facilities. In other words, individual observations within each facility are viewed as "within-center attributes", with the multiple facilities providing the independent information to estimate treatment effects (both within- and between-facilities).

2.2. Social Network Engeneity and U-Statistics--A Match Made in Heaven

The mainstream analytic paradigm is replete with applications of traditional statistical models that fail to recognize the social-network induced endogeneity, or when they do, they fail to address this critical issue. Some simply ignore endogeneity and analyze social network data using standard methods such as regression models (Centola, 2011; 2011). While others apply specialized social network models attempting to address this issue, but are inadvertently continuing this flaw, since they are still confined to modeling principles underlying standard statistical models (El-Sayed et al., 2012; Borgatti et al., 2002; Robins and Morris, 2007; Snijders and Borgatti, 1999; Snijders et al., 2002; Strauss and Ikeda, 1990; Wasserman and Faust, 1994). For example, a popular approach for addressing social-network induced endogeneity is to use resampling methods such as the Jackknife and Bootstrap. Although viable options for estimating variability of estimates in complex models involving inter-dependent variables, they lack the ability to address between-subject interactions, yielding underestimated standard errors (Lu et al., 2013).

The above analysis shows that predominant statistical paradigm for modeling within-subject attributes is fundamentally flawed when applied to modeling social network structures. To address the social network endogeneity, we must break the confines of such conventional thinking and model relationships between attributes from different subjects. For example, consider the likelihood of having a tie, or relationship, between two subjects in a social network, an important feature of the social network known as the density. If the social network has n subjects and f_{ij} denotes the indicator of a tie between any ith and jth subjects from the network, i.e., $f_{ij} = 1$ if a tie is present and $f_{ij} = 0$ otherwise, then the density of the social network is the probability $\theta = \Pr(f_{ij} = 1)$.

To estimate this density parameter, we may calculate the proportion of connected pairs to all possible pairs and then calculate the standard error to estimate the variability of the estimate so we can construct confidence intervals or test hypotheses concerning the value of the density. Although the first task is quite straightforward, the second is not. For example, an unbiased estimate of θ is a proportion: $\hat{\theta}_n = \binom{n}{2}^{-1} \sum_{(i,j)\in C_2^n} f_{ij}$, where $C_2^n = \{(i,j); 1 \le i < j \le n\}$ denotes all distinct combinations of (i, j) from the integer set $\{1,2,\ldots,n\}$. However, standard methods for a binomial proportion or even methods specialized for social network analysis like the exponential random graph model will not provide correct standard error estimates of $\hat{\theta}_n$ (Lu et al., 2013). Even resampling methods such as Jackknife and Bootstrap will not help either (Lu et al., 2013). The problem is again the endogeneity in the outcomes f_{ij}. Since two subjects may be potentially connected, all the f_{ij} are dependent. For example, f_{ij}, f_{ik} and f_{jk} are all dependent, since f_{ij} and f_{ik} involve the ith subject, f_{ij} and f_{ik} contains the jth subject, while f_{ik} and f_{jk} include the kth subject. If ignored as by any of the aforementioned methods, inference about the density θ gives rise to biased inference (Lu et al., 2013).

The theory of U-statistics is uniquely positioned to address the between-subject endogeneity in social network analysis (Hoeffding, 1948; Kowalski and Tu, 2007). Classic examples of U-statistics include the Mann-Whitney-Wilcoxon rank-sum and signed-rank tests and Kendall's tau (Kowalski and Tu, 2007). More recently, this class of statistics has also been extended to provide inference about correlations within a longitudinal data setting (Ma et al., 2008; 2010 and 2011; Tu et al., 2007). The key difference between U-statistics and conventional statistics is that unlike the latter the former class of statistics is defined by between-subject attributes. For example, within the context of social network density, the unbiased estimate, $\hat{\theta}_n = \binom{n}{2}^{-1} \sum_{(i,j)\in C_2^n} f_{ij}$, is a U-statistic, defined by the connection indicator f_{ij} among all pairs of subjects in the network. Since the theory of U-statistics addresses the endogeneity of f_{ij}, it provides correct standard error estimates and thereby yields valid inference about θ (Lu et al., 2013).

2.3. Functional Response Models for Integrating between- and within-Subject Attributes

Although the U-statistics theory may be used to facilitate inference about the social network density, its applications are limited and may not be applied to more complex models. For example, if we are interested in estimating the mean of a measure of some within-subject attribute of interest such as depressive symptoms within the context of social networks, we know that standard methods will again underestimate the sampling variability of the estimate because of potential correlations between subjects with social network ties. To account for the correlation induced by the social network endogeneity, we can model such within-subject characteristics together with the between-subject attributes such as the social network density. The functional response models (FRM) is uniquely positioned to tackle the complexity of such a system of relationships involving measures of both between- and within-subject attributes.

The FRM integrates U-statistics and distribution-free (semi-parametric) models to provide a flexible framework for modeling complex relationships involving between-subject dynamics and high-order moments based on responses from multiple subjects (Kowalski and Tu, 2007). Existing regression models are defined based on a single-subject response, severely limiting their applications in practice. For example, in the linear model in Equation (1), the dependent variable is a single-subject response y_i. Although modern regression models such as the generalized linear models and even non-linear models have more complex forms, the model specification still only involves a single subject response.

For example, the logistic regression, a member of generalized linear models, is defined by:

$$y_i \sim Bern(\mu_i), \quad \text{logit}(\mu_i) = \beta_0 + \beta_1 x_i, \tag{3}$$

where y_i is a binary variable, x_i is a predictor, $Bern(\mu)$ denotes the Bernoulli distribution with mean μ and $\text{logit}(\mu) = \log\left(\frac{\mu}{1-\mu}\right)$ is the logit link (Tang et al., 2012). Although the conditional mean μ_i of y_i given x_i is not set equal to the linear predictor $\eta_i = \beta_0 + \beta_1 x_i$ as in linear regression, the dependent variable is a single subject linear response y_i.

Many quantities of interest require modeling more complex distributions and response functions. For example, the error term ε_y in the linear model in Equation (1) is assumed to follow the normal distribution, which is often at odds with the real data distribution in practice. The distribution-free, or semi-parametric, regression approach addresses this flaw by modeling only the conditional mean:

$$E(y_i \mid x_i) = \beta_0 + \beta_1 x_i, \tag{4}$$

and uses estimating equations for inference about the vector of model parameters $\beta = (\beta_0, \beta_1)^T$, where the superscript ^{T} denotes the transpose of a vector. Without imposing any mathematical distribution, the distribution-free linear regression in Equation (4) provides valid inference for a wider class of distributions.

Although providing more robust inference, the distribution-free models are still confined to the single-subject based linear response y_i. Many modern applications require more not only complex functions, but also multi-subject responses. For example, when addressing overdispersion in modeling count data using the negative binomial distribution, it is necessary to model y_i^2 in addition to the linear response y_i. Within our current context, we need to model between-subject attributes such as connections between two subjects for the density as discussed in Section 2.2.

The FRM integrates U-statistics to address the limitations of distribution-free regression models by allowing for complex response functions involving multiple-subject responses:

$$E\left[\mathbf{f}\left(y_{i_1},\ldots,y_{i_q}\right)\mid \boldsymbol{x}_{i_1},\ldots,\boldsymbol{x}_{i_q}\right]=\mathbf{g}\left(\boldsymbol{x}_{i_1},\ldots,\boldsymbol{x}_{i_q};\boldsymbol{\theta}\right),\quad \left(i_1,\ldots,i_q\right)\in C_q^n, \tag{5}$$

where $\mathbf{f}(\cdot)$ is some function, $\mathbf{g}(\cdot)$ is some smooth function (with continuous second-order derivatives), C_q^n is the set of distinct q-element combinations $(i_1,\ldots,i_q)$ from the integer set $\{1,\ldots,n\}$, and $\boldsymbol{\theta}$ is a vector of parameters. By generalizing the response variable in this fashion, the FRM greatly broadens applications of regression models to address a range of methodological issues such as mixture models (Yu et al., in press), instrumentation (Ma et al., 2008; 2010; 2011), SEM-based longitudinal mediation analysis (Gunzler et al., in press) and social networking (Yu et al., 2011).

Within our context of social network analysis, the FRM is readily applied to extend the density for a homogeneous population discussed in Section 2.2 to a regression setting to accommodate heterogeneity within the study population. For example, if we set $q = 2$ in the FRM in Equation (5), we obtain:

$$E\left[f\left(y_i,y_j\right)\mid \boldsymbol{x}_i,\boldsymbol{x}_j\right]=g\left(\boldsymbol{x}_i,\boldsymbol{x}_j;\boldsymbol{\theta}\right),\quad \left(i,j\right)\in C_2^n. \tag{6}$$

If $f(y_i, y_j)$ defines the connection between the ith and jth subjects, i.e., $f(y_i, y_j) = 1$ if the two subjects have a tie and $f(y_i, y_j) = 0$ otherwise, then $g(\boldsymbol{x}_i, \boldsymbol{x}_j;\boldsymbol{\theta})$ is the mean of the social network connection, or density, within the subgroups defined by the covariates $\boldsymbol{x}_i$ and $\boldsymbol{x}_j$. Note that y_i and y_j in the density application do not represent any real within-subject attribute and simply serve as "dummy" indicators of the two different subjects within the social network. The advantage of using FRM to model the social network density is the inclusion of covariates $\boldsymbol{x}_i$ and $\boldsymbol{x}_j$ to allow the density to vary as a function of within-subject characteristics, akin to standard regression models for within-subject attributes.

For example, in research on homophily such as peer socialization, we need to model the tendency of individuals to associate (between-subject attributes) with similar others (within-subject attributes). In such applications, the between-subject response function, $f(y_i, y_j)$, is linked to the within-subject covariates $\boldsymbol{x}_i$. As a simple illustration, consider a single individual attribute depression, i.e., $\boldsymbol{x}_i = 1$ if the subject has depression and $\boldsymbol{x}_i = 0$ otherwise. If $f(y_i, y_j)$ is a binary indicator of a tie between the ith and jth subjects, then $g(\boldsymbol{x}_i, \boldsymbol{x}_j;\boldsymbol{\beta})$ in (6) is the probability of a tie between these two subject. Thus, as in logistic regression, we model $g(\boldsymbol{x}_i, \boldsymbol{x}_j;\boldsymbol{\beta})$ as:

$$\operatorname{logit}\left(g\left(x_i,x_j;\boldsymbol{\beta}\right)\right)=\exp\left(\beta_0+\beta_1\left|x_i-x_j\right|+\beta_2x_ix_j\right),\quad \left(i,j\right)\in C_2^n,$$

where $\operatorname{logit}(u)=\frac{u}{1-u}$ denotes the logit function (Tang et al., 2012). Under this model, we have:

$$g\left(x_i, x_j; \boldsymbol{\theta}\right) = \begin{cases} \theta_1 & \text{if both subjects are depressed} \\ \theta_2 & \text{if both subjects are non-depressed} \\ \theta_3 & \text{if one is depressed and the other is not} \end{cases},$$

where $\theta_1 = \frac{\exp(\beta_0)}{1+\exp(\beta_0)}$, $\theta_2 = \frac{\exp(\beta_0+\beta_2)}{1+\exp(\beta_0+\beta_2)}$ and $\theta_3 = \frac{\exp(\beta_0+\beta_1)}{1+\exp(\beta_0+\beta_1)}$, denoting the probabilities of connection between two two-depressed, one depressed and one non-depressed and two non-depressed subjects.

We can use the FRM to test hypotheses concerning the strength of ties both between- and within-groups of depressed and non-depressed subjects. For example, the null hypothesis $H_0{:}\theta_1 = \theta_2$ tests whether the probability of social connection between two depressed people is the same as that between two non-depressed individuals, while the null $H_0{:}\theta_1 = \theta_3$ examines whether a depressed person is more (or less) likely to have a relationship between a depressed than non-depressed peer.

As noted earlier, standard statistical models generally provide correct model estimates, but invalid inference because of the social network endogeneity. For example, within the context of the FRM in (6), let y_i denotes another measure of a within-subject attribute such as suicide ideation (Heisel et al., 2009). Then, the difference in the mean of y_i between the two groups defined by the depression indicator x_i can be modeled by a linear regression, $E\left(y_i \mid x_i\right) = \gamma_0 + x_i\gamma_1$. Standard procedures such as the maximum likelihood or estimating equations still provide good estimates. However, standard inference procedures based on the independence of y_i such as the maximum likelihood or even estimating equations underestimates the variability of such estimates of $\boldsymbol{\gamma} = (\gamma_0, \gamma_1)'$, because y_i are not independent.

By extending the FRM in (5) to simultaneously model this measure of within-subject attribute and the between-subject social network ties, we will be able to provide valid inference about the population of interest. For example, let $\delta_{ij} = 1$ if the ith and jth subjects have a tie and $\delta_{ij} = 0$ otherwise. We then define our FRM as follows:

$$E\left[f_1\left(y_i, y_j\right) \mid x_i, x_j\right] = g_1\left(x_i, x_j; \boldsymbol{\theta}\right), \quad E\left[f_2\left(y_i, y_j\right) \mid x_i, x_j\right] = g_2\left(x_i, x_j; \boldsymbol{\theta}\right) \quad (7)$$

where

$$f_1\left(y_i, y_j\right) = \delta_{ij}, \quad f_2\left(y_i, y_j\right) = y_i + y_j, \quad \boldsymbol{\theta} = \left(\boldsymbol{\beta}', \boldsymbol{\gamma}'\right)',$$

$$\operatorname{logit}\left(g_1\left(x_i, x_j; \boldsymbol{\theta}\right)\right) = \beta_0 + \beta_1\left|x_i - x_j\right| + \beta_2 x_i x_j, \quad g_2\left(x_i, x_j; \boldsymbol{\theta}\right) = 2\gamma_0 + \left(x_i + x_j\right)\gamma_1.$$

The FRM in (7) extends the FRM in (6) to include an additional component to model the measure of the within-subject attribute y_i, defined by the function response $f_2(y_i, y_j)$ and the associated mean $g_2(x_i, x_j; \boldsymbol{\theta})$. The parameters $\boldsymbol{\gamma} = (\gamma_0, \gamma_1)'$ has the same interpretation as in the standard regression model $E\left(y_i \mid x_i\right) = \gamma_0 + x_i\gamma_1$. The alternative expression is used to model the within-subject attribute measure together with the between-subject social network tie. Because of this joint modeling, the FRM in (7) provides valid inference about $\gamma = (\gamma_0, \gamma_1)'$.

3. Issues on Causal Inference

As in classic statistical analysis, the issue of causation is also at the heart of social network analysis. In fact, it is this concept of causality that has continued to drive research and interest in understanding the roles that social networks play in the cause of a wide range of phenomena known as adoption and diffusion of innovation, such as the rapid transition of power in a country, success rates in job search and product marketing and diffusion of prosocial norms and behaviors in distressed neighborhoods. An extensive literature has emerged to attempt to understand the cause and its mechanism using the concept of disease contagion as the basic model. A number of approaches have been proposed to create a theoretical framework to tease out *contagion* from *convergence*.

For example, in a study on peer socialization and its effect on adolescent's depression, Kiuru et al., (2012) indicated that the current premise used for studying this causal relationship was flawed, because they attributed increases in depression in the adolescents who had depressed peers to contagion without considering any alternative possibilities. They argued that these studies conflated measures of contagion with measures of convergence, two fundamentally different concepts. Under contagion, higher levels of depressive symptoms among a networked group of adolescents are caused by the contagious effect of one or more depressed adolescents. However, the increased levels of depressive symptoms may also be the result of convergence or homophily, i.e., adolescents simply bond with similar others.

Such a debate between causation (contagion) and association (convergence or homophily) is not new and in fact, has been extensively researched, forming an impressive body of literature known as causal inference. Although the concepts of causal inference apply to social network data, the models and methods do not, because of the social network endogeneity discussed in Section 2. Below, we briefly review the concept of counterfactual outcome, the underpinning of the causal inference enterprise, and the theoretical framework for causal inference based on this important concept.

3.1. The Potential Outcome Concept

Although the concept of causal treatment effect is straightforward, a formal definition is actually quite complex. A popular and accepted one is the counterfactual outcome based causal inference paradigm. Originally introduced by Fisher (1918) in the context of inference for randomized experiments and later extended by Rubin (1974) to non-randomized studies as well as randomized controlled trials, a fundamental building block of the causal inference paradigm is the concept of *potential outcomes*. A subject's causal effect is the comparison of the potential outcomes.

Consider a study with a treatment and control condition. For a variable (within-subject attribute) of interest such as depressive symptoms, each subject has two potential outcomes associated with the treatment and control. The potential outcomes are counterfactual, since each subject is only assigned to one treatment condition and thus only the one associated with the assigned treatment is observed.

With the concept of potential outcomes, we can truly speak of causal effect. If the two potential outcomes for each subject are y_{i1} and y_{i0} for the treated and control condition, respectively, the difference between the two $\Delta i = y_{i1} - y_{i0}$ is the treatment effect for the subject. Since this difference is based on the outcomes from the same subject in response to the two treatments, it is free of any selection bias and thus is the effect of the treated condition. Unfortunately, since only one of the potential outcomes is observed, this difference is not computable, ruling out direct applications of standard statistical models. A large part of the causal inference literature centers on how to estimate the mean, or population-level, causal treatment effect, $\Delta = E(y_{i1} - y_{i0}) = E(y_{i1}) - E(y_{i0})$, without observing both outcomes y_{i1} and y_{i0} for each subject, where $E(\cdot)$ denotes the mathematical expectation.

3.2. Causal Models for within-Subject Attributes

For randomized studies, treatment assignment is independent of the potential outcomes. In this case, every subject selected is a representative of the study population and thus the mean $E(y_{i1})$ ($E(y_{i0})$) is the same for everyone assigned to the treated (control) condition, i.e., $E(y_{i1}) = (E(y_{j1})$ $(E(y_{i0}) = (E(y_{j0}))$. This means that we can estimate the mean of the treated $E(y_{i1})$ and control ($E(y_{i0})$) based on the different subjects from the treated and control groups using standard methods such as the sample mean, $\frac{1}{n_k}\sum_{i=1}^{n_k} y_{ik}$, for the treated ($k = 1$) and control ($k = 0$) group, where n_k denotes the sample size of the kth treatment condition. For non-randomized trials such as observational studies in epidemiologic research, the subjects selected are generally not representative of the study population and estimates such as the sample mean based on the observed data generally do not represent the effect of treatment, or exposure. A number of approaches are available to address such selection bias in observational studies.

To illustrate such methods, let us consider the problem of contagion of depression again. To see if depression is contagious in a population of interest, we would ideally conduct a randomized controlled study. In the treated condition, every subject would receive an intervention consisting of socializing with a depressed subject, while in the control condition, each subject would be deprived any opportunity of socialization with a depressed individual. Based on the counterfactual outcome paradigm, a significant difference between the means of the depression outcome from the two treatment groups would support the contagion of depression hypothesis.

However, such a randomized controlled study would be difficult to carry out, since it is neither practical nor ethical to force people to socialize with depressed individuals or deprive people of opportunities to have contact with their depressed friends. Without the ability to manipulate the treatment condition, differences observed in depression symptoms between people with and without depressed friends generally do not implicate causation. Higher symptom levels among those with depressed friends could simply be the result of the convergence, or homophily, phenomenon, i.e., people tend to initiate friendship with similar peers. To inference causality from such empirical differences, we can use a number of causal methods.

The case-control design is a classic method for ascertaining causal relationships in non-randomized observational studies (Tang et al., 2012). In a case-control study on examining the relationship between some exposure variable and disease of interest, we first select a sample from a population of diseased subjects, or cases (people with depressed friends within our context). Such a case sample is usually retrospectively identified. We then randomly select a sample of disease-free individuals, or controls, from a non-diseased population (subjects without depressed friends in our context), with identical or similar characteristics believed to predispose subjects to the disease. Since the cases and controls are closely matched to each other in all predisposed conditions for the disease except for the exposure status, differences between the case and control groups may be attributable to the effect of exposure.

For the case-control design to work well, we must be able to find good controls of the cases. If $\boldsymbol{x}_i$ denotes the set of covariates for matching cases and controls, we must pair each case and control with identical or similar $\boldsymbol{x}_i$'s. For example, if $\boldsymbol{x}_i$ consists of age, gender and patterns of smoking (e.g., frequency and years of smoking), we may try to pair each lung cancer patient with a healthy control, having same gender, same (or similar) age and smoking patterns. However, as the dimension of $\boldsymbol{x}_i$ (number of within-subject attributes) increases, it becomes more difficult to match cases with controls.

One popular approach for addressing the high dimensional issue is the *propensity score* (PS) method (Rosenbaum and Rubin, 1983). The PS models the probability, or propensity score, of exposure status (case vs. control) based on $\boldsymbol{x}_i$ using models for binary outcomes such as logistic regression, and match subjects with same or similar propensity scores to create strata. Differences between the cases and control subjects within each stratum are computed and integrated across all strata to create treatment (exposure) effects.

Although simple in principle, PS has a number of drawbacks when implemented in practice. Since the covariates $\boldsymbol{x}_i$ often contain continuous variables such as age or measures of some medical or mental health conditions and severities in real studies, it is not possible to match cases and controls with exactly the same propensity score. Thus, in real studies subjects are matched with similar scores instead. This limitation in matching subjects not only fails to completely remove selection bias, but also yields strata based on subjective decisions (Rosenbaum and Rubin, 1984). The PS also lacks nice properties of formal statistical models such as asymptotic properties of estimates.

A popular alternative is the marginal structural model (MSM) (Hernan et al., 2002). Unlike PS, MSM uses the propensity score as a weight to adjust for the bias in treatment selection, a technique known as the inverse probability weighting (IPW) and first used in survey research to address selection bias in households selectively sampled from a targeted region of interest (Tang et al., 2012). By employing the IPW technique to address treatment selection, MSM not only completely removes selection bias, but also yields estimates of causal treatment effect with nice asymptotic properties. In other words, unlike PS, the accuracy of inference about treatment effects under MSM continues to improve as the sample size increases. Further, MSM is readily applied to longitudinal data with missing values.

Under MSM, we model the propensity score $\pi(\boldsymbol{x}_i)$ as a function of the covariates $\boldsymbol{x}_i$ using models for binary response such as logistic regression. The mean $E(y_{ik})$ for the kth group is estimated by a weighted sample mean, $\frac{1}{n_k}\sum_{i=1}^{n_k}\frac{1}{\pi(\boldsymbol{x}_i)}y_{ik}$, rather than the usual sample mean, to account for selection probabilities across the different subjects. Since the propensity score is used as a weight to counter-effect the selection probability of each subject, rather than a criterion to create a group of subjects with similar selection probabilities as in PS, there is no residual selection bias in the MSM estimate of causal effect.

In practice, we may not collect all the information that dictates the selection process in an observational study, because of the retrospective nature of such studies. Many times, this may simply be the result of limited measurement capabilities. For example, if certain genetic traits that predispose a subject to the development of depression could not be measured, the propensity score would not be able to capture this source of selection bias and the PS or even the MSM would still yield biased estimates. Although such hidden bias is impossible to account for, its effect on model estimates (e.g., weighted sample means) can be assessed by conducting sensitivity analysis (Rosenbaum and Rubin, 1983; Scharfstein et al., 1999; Tang et al., 2012). In such analysis, we include some latent variable b_i in the model for the propensity score to represent the contribution of hidden bias and use the resulting $\pi(\boldsymbol{x}_i, b_i)$ as the propensity score for each subject. By assessing how sensitive PS or MSM estimates are in response to the presence of hidden bias, we can get some ideas about the reliability of the estimates of causal effect obtained under the model based on the original propensity score $\pi(\boldsymbol{x}_i)$.

As with statistical inference, existing causal models and inference approaches are not directly applicable to social network data because of the social network endogeneity. However, by integrating the concept of potential outcome with the FRM as we illustrated in Section 2.3 in modeling both within- and between-subject attributes, in principle we will be able to address the correlations induced by the social network endogeneity among the outcomes of different subjects. Much work is needed to extend the potential outcome based causal models to the context of FRM to develop a coherent framework for modeling causal relationships between different within-subject attributes (such as depressive symptoms and friendship with a depressed individual) within the context of social network data.

4. Discussion

We discussed social network endogeneity, a defining feature of social network data, and its implications for statistical and causal inferences when modeling such data using conventional standard statistical and causal models. Unlike the concept of endogeneity in standard statistical analysis, the social network engeneity is the result of reciprocal relations between subjects, rather than similar relationships between different within-subject attributes. Although this conceptual difference may seem minor in appearance, this shift from within- to between-subject attributes is so significant that a fundamentally new approach is needed to model social network structures and constructs such as the social network density as illustrated in this paper.

The influential social-ecological framework posits that individuals are nested in social systems that directly influence individual development (Bronfenbrenner, 1979). Over the past twenty years, recognition of such clustered data, or reciprocal relations among within-subject attributes, has led to the wide applications of multi-level and structural equation models to address resulting endogeneity, or interdependence among variables. Just as addressing flawed estimates and/or standard errors has become standard practice for such endogenous relationships between variables, so ought practices for recognizing and addressing the social network endogeneity, or links between subjects. Given the wide dissemination of social network analysis, addressing fundamental flaws in standard statistical analysis when applied to social network data is a key direction for sciences (that draw on such methods) to grow with integrity.

For instance, in the past decade, systems science, that includes social network analyses, has grown to become a leading strategic direction in health and public health sciences. The National Institutes of Health continue to sponsor annual summer training institutes in systems science methods. This intensive training program includes a track on social network analysis. The interest of such leading health sciences funding agencies is reinforced by rising interest in studying `virtual' environments (e.g., in prevention science) given newly available social media data applicable for a host of fields (Centola, 2010; Cho et al., 2011; Sadilek et al., 2012). Such trends suggest that inquiry into valid statistical inference of social network data is a key pressing direction to support the growth of new scientific directions that will continue to gain momentum in the immediate future.

ACKNOWLEDGMENTS

We are very grateful to Dr. Vincent Silenzio, Dr. Christopher Homan and Dr. Wan Tang for their constructive comments about random graphs. We wish to thank Dr. Eric Caine and the "Scholarly Pens" writing group, both of the Department of Psychiatry, for their constructive feedback as readers. We also wish to thank Jessica Poweski for her assistance for preparing this manuscript for submission.

REFERENCES

[1] Akpan-Obong, P. & Parmentier, M. J. C. (2013). *Hash tags, status updates and revolutions: A comparative analysis of social networking in political mobilization*. In Social Networking: Recent Trends, Emerging Issues and Future Outlook, edited by Lu, N., White, A.M. and Tu, X.M., NOVA Science, NY.

[2] Angrist, J. D. & Krueger, A. B. (2001). Instrumental variables and the search for identification: From supply and demand to natural experiments. *Journal of Economic Perspectives*, *15*, 69-85.

[3] Archer, J., Bower, P., Gilbody, S., Lovell, K., Richards, D., Gask, L., Dickens, C. & Coventry, P. (2012). *Collaborative care for depression and anxiety problems*. Wiley, New York.

[4] Bollen, K. A. (1989). *Structural Equations with Latent Variables. Wiley, New York.*

[5] Borgatti, S. P., Everett, M. G. & Freeman, L. C. (2002). *Ucinet for Windows: Software for social network analysis*. Harvard, MA: Analytic Technologies.

[6] Bronfenbrenner, U. (1979). *The ecology of human development: Experiments by nature and design*. Harvard University Press.

[7] Centola, D. (2010). The spread of behavior in an online social network experiment. *Science*, *329*, 1194-1197. DOI: 10.1126/science.1185231.

[8] Centola, D. (2011). An experimental study of homophily in the adoption of health behavior. *Science*, *334*, 1269-1272. DOI: 10.1126/science.1207055.

[9] Cho, E., Myers, S. A. & Leskovec, J. (2011). *Friendship and mobility: User movement in location-based social networks*. ACM SIGKDD International Conference on Knowledge Discovery and Data Mining (KDD).

[10] El-Sayed, A. M., Scarborough, P., Seemann, L. & Galea, S. (2012). Social network analysis and agent based modeling in social epidemiology. *Epidemiologic Perspectives & Innovations*, *9*, 1-9.

[11] Feinberg, M. E., Riggs, N. R. & Greenberg, M. T. (2005). Social networks and community prevention coalitions. *J Prim Prev.*, *26*(4), 279-298.

[12] Fisher, R. A. (1918). The causes of human variability. *Eugenics Rev*. 10, 213-20.

[13] Goodreau, S. M., Handcock, M. S., Hunter, D. R., Butts, C. T. & Morris, M. (2008). A STATNET Tutorial. *J Stat Softw*, *24*(9), 1-27.

[14] Gunzler, D., Lu, N., Tang, W. & Tu, X. M. *A class of distribution-free models for longitudinal mediation analysis*, Psychometrika, in press.

[15] Heisel, M. J., Duberstein, P. R., Talbot, N. L., King, D. A. & Tu, X. M. (2009). Adapting interpersonal psychotherapy for older adults at risk for suicide: Preliminary findings. *Professional Psychology: Research and Practice*, *40*(2), 156-164.

[16] Hoeffding, W. (1948). A class of statistics with asymptotically normal distribution. *Annals of Mathematical* Statistics, *19*, 293-325.

[17] Kellam, S. G. (2012). Developing and maintaining partnerships as the foundation of implementation and implementation science: reflections over a half century. *Adm Policy Ment Health*, *39*(4), 317-320. doi:10.1007/s10488-011-0402-8

[18] Kiuru, N., Burk, W. J., Laursen, B., Nurmi, J. & Salmela-Aro, K. (2012). Is depression contagious? A test of alternative peer socialization mechanisms of depressive symptoms in adolescent peer networks. *Journal of Adolescent Health*, *50*, 250–255.

[19] Kowalski, J. & Tu, X. M. (2007). *Modern Applied U Statistics*. Wiley, New York.

[20] Lakon, C. M., Hipp, J. R. & Timberlake, D. S. (2010). The social context of adolescent smoking: A systems perspective. *Am J Public Health*, *100*(7), 1218-1228. doi: 10.2105/AJPH.2009.167973.

[21] Lu, N., White, A. M., Wu, P., He, H., Hu, J., Feng, C. & Tu, X. M. (2012). *On statistical inference for social network density*. Technical Report, Department of Biostatistics and Computational Biology, University of Rochester, Rochester, New York.

[22] Luke, D. A., Harris, J. K., Shelton, S., Allen, P., Carothers, B. J. & Mueller, N. B. (2010). Systems analysis of collaboration in 5 national tobacco control networks. *Am J Public Health*, *100*(7), 1290-1297. doi: 10.2105/AJPH.2009.184358

[23] Ma, Y., Tang, W., Feng, C. & Tu, X. M. (2008). Inference for Kappas for longitudinal study data: Applications to sexual health research. *Biometrics*, *64*, 781-789.

[24] Ma, Y., Tang, W. & Tu, X. M. (2010). Modeling Concordance Correlation Coefficient for longitudinal study data. *Psychometrika*, *75*, 99-119.

[25] Ma, Y., Tang, W. & Tu, X. M. (2011). Modeling Cronbach Coefficient Alpha for longitudinal study data. *Statistics in Medicine.*, *29*(6), 659-670.

[26] Muthen and Muthen (2012). Mplus Version 7, California.

[27] Palinkas, L. A., Holloway, I. W., Rice, E., Fuentes, D., Wu, Q. & Chamberlain, P. (2011). Social networks and implementation of evidence based practices in public youth-serving systems: a mixed-methods study. *Implement Sci.*, *6*(113), 1-11.

[28] Raghupathi, V. & Fogel, J. (2013). Opinion leadership: The role of opinion leaders and internet marketing through social networking websites. *In Social Networking: Recent Trends, Emerging Issues and Future Outlook*, edited by Lu, N., White, A.M. and Tu, X.M., NOVA Science, NY.

[29] Ramirez-Ortiz, G., Caballero-Hoyos, R., Ramirez-Lez, G. & Valente, T. W. (2012). The effects of social networks on tobacco use among high-school adolescents in Mexico. *Salud Plica De Mico.*, *54*(4), 433-441.

[30] Robins, G. & Morris, M. (2007). Advances in exponential random graph (p*) models. *Soc Networks.*, *29*(2), 169-172.

[31] Rosenbaum, P. & Rubin, D. (1983). The central role of the propensity score in observational studies for causal effects. *Biometrika*, *70*, 41-55.

[32] Rosenbaum, P. R. & Rubin, D. B. (1984). Reducing bias in observational studies using subclassification on the propensity score. *Journal of the American Statistical Association*, *79*, 516-24.

[33] Rubin, D. B. (1974). Estimating causal effects of treatments in randomized and non-randomized studies. *J. Educ. Psychol.*, *66*, 688-701.

[34] Sadilek, A., Kautz, H. & Silenzio, V. (2012). *Modeling spread of disease from social interactions.* In Sixth AAAI International Conference on Weblogs and Social Media (ICWSM).

[35] Scharfstein, D. O., Rotnitzky, A. & Robins, J. M. (1999). Adjusting for nonignorable Drop-out using semi-parametric non-response models. *Journal of the American Statistical Association*, *94*, 1096-1146.

[36] Snijders, T. & Borgatti, S. (1999). Non-Parametric standard errors and Tests for network statistics. *Connections*, *22*(2), 161-170.

[37] Snijders, T. A. (2002). Markov Chain Monte Carlo estimation of exponential random graph models. *Journal of Social Structure*, *3*(2), 1-40.

[38] Strauss, D. & Ikeda, M. (1990). Pseudolikelihood estimation for social networks. *J Am Stat Assoc.*, *85*(409), 204-212.

[39] Tang, W., He, H. & Tu, X. M. (2012). *Applied Categorical and Count Data Analysis.* Chapman & Hall/CRC, FL.

[40] Trimble, L. B., Kmec, J. A. & McDonald, S. (2013). *Social networks and the job search: focusing on people asked to provide job assistance*. In Social Networking: Recent Trends, Emerging Issues and Future Outlook, edited by Lu, N., White, A.M. and Tu, X.M., NOVA Science, NY.

[41] Tu, X. M., Feng, C., Kowalski, J., Tang, W., Wang, H., Wan, C. & Ma, Y. (2007). Correlation analysis for longitudinal data: applications to HIV and psychosocial research. *Statistics in Medicine*, *26*, 4116–4138.

[42] Valente, T. W., Chou, C. P. & Pentz, M. A. (2007). Community coalitions as a system: effects of network change on adoption of evidence-based substance abuse prevention. *Am J Public Health*, *97*(5), 880-886.

[43] Wasserman, S. & Faust, L. (1994). *Social Network Analysis: Methods and Applications*. Cambridge University Press.

[44] Yu, Q., Tang, W., Kowalski, J. & Tu, X. M. (2011). Multivariate U-Statistics: A tutorial with applications. *Wiley Interdisciplinary Reviews – Computational Statistics*, *3*, 457-471. doi: 10.1002/wics.178.

[45] Yu, Q., Chen, R., Tang, W., He, H., Gallop, R., Crits-Christoph, P., Hu, J. & Tu, X. M. (2013). Distribution-free models for longitudinal count responses with over-dispersion and structural zeros, *Statistics in Medicine*, in press.

INDEX

A

adolescents, 98, 101, 102, 103, 104, 105, 106, 107, 108, 111, 112, 176, 182
ageing, 55, 56, 58, 59, 65, 69
association network, 3, 13
association strength, 3, 5, 9, 12, 14

B

best potential friend, 113, 114
Bolivia, 21, 23, 26, 29, 32, 33, 34, 35, 36, 38

C

causal inference, viii, 167, 168, 169, 176, 177, 179
centrality, 7, 8, 11, 12, 13, 15, 41, 49, 50, 51, 63, 64, 65, 66, 128, 131, 133, 134, 137
climate change, 55, 56, 57, 58, 59, 60, 61, 62, 65, 66, 67, 69, 70, 71
clustered data, 167, 170, 171, 179
cohesion, 5, 15, 16, 58
contagion, 98, 167, 168, 170, 176, 177

D

distribution-free models, 173, 181

E

employment, 139, 140, 141, 142, 143, 146, 147, 149, 154, 156, 157, 158, 160, 161, 162

F

facebook, 97
functional response models, 167, 173

H

haemophilia, 101, 102, 103, 104, 105, 106, 107, 108, 109, 110, 111, 112
homophily, 83, 86, 89, 91, 167, 174, 176, 177, 181

I

internet, 38, 60, 91, 97, 103, 106, 109, 136, 162, 182

L

literature review, viii, 134

M

marginal structural models, 167

N

natural language processing, 89
network contacts, 139, 140, 141, 142, 143, 144, 145, 156, 161, 162
network structure, 1, 3, 5, 9, 10, 12, 86
networks), 65
Nigeria, 4, 16, 17, 18, 21, 23, 26, 29, 30, 31, 33, 34, 35, 36, 37, 38, 39

O

opinion leaders, viii, 125, 127, 128, 129, 130, 131, 132, 133, 134, 135, 136, 137, 138, 182

P

PageRank, 113, 114, 115, 116, 117, 118, 119, 120, 121, 122, 132
personality recognition, 41, 43, 44, 45, 49, 51
political mobilization, 21, 23, 26, 28, 32, 33, 36, 37, 180
potential outcome, 167, 176, 177, 179
preferential association, 1, 3, 5, 9, 10, 11, 12, 14, 15
propensity score, 167, 178, 179, 182
provision of help with a job search, 153

R

rural Bangladesh, 55, 56, 57, 59, 60, 61, 68, 70

S

social media, vii, 26, 28, 30, 32, 35, 53, 74, 75, 78, 92, 93, 95, 98, 109, 125, 127, 129, 136, 168, 180
social network density, 167, 172, 174, 179, 181
social networking services, 73
stochastic independence, 167

U

U-statistics, 167, 172, 173, 174

W

word-of-mouth, 125, 127, 128, 131, 133, 134, 135, 136, 137, 138